THE MASS SPECTROMETER

THE WYKEHAM SCIENCE SERIES

General Editors:

PROFESSOR SIR NEVILL MOTT, F.R.S.
Emeritus Cavendish Professor of Physics
University of Cambridge

G. R. NOAKES
Formerly Senior Physics Master
Uppingham School

Chemisty Editor:

M. P. BERRY
Senior Teacher, Science
Chislehurst and Sidcup Grammar School

The aim of the Wykeham Science Series is to introduce the present state of the many fields of study within science to students approaching or starting their careers in University, Polytechnic, or College of Technology. Each book seeks to reinforce the link between school and higher education, and the main author, a distinguished worker or teacher in the field, is assisted by an experienced sixth form schoolmaster.

THE MASS SPECTROMETER

J. R. Majer
Chemistry Department,
University of Birmingham

WYKEHAM PUBLICATIONS (LONDON) LTD
(A member of the Taylor & Francis Group)
CRANE, RUSSAK & COMPANY INC.
1977

Sole Distributors for the United States of America and Canada
CRANE, RUSSAK & COMPANY, INC. NEW YORK

First published 1977 by Wykeham Publications (London) Ltd.

© 1977 J. R. Majer. All rights reserved. No part of this publication may be reproduced, stored in a retrieval system, or transmitted, in any form or by any means, electronic, mechanical, photocopying, recording or otherwise, without the prior permission of the copyright owner.

ISBN 0 8448 1171 8

Library of Congress Catalog Card Number 77-15307

Printed in Great Britain by Taylor & Francis (Printers) Ltd. Rankine Road, Basingstoke, Hants RG24 0PR.

CONTENTS

Preface   vii

Terms used in mass spectrometry   ix

Chapter 1   PRINCIPLES OF MASS SPECTROMETRY

1.1 Development of the mass spectrometer	1
1.2 Commercial mass spectrometers	14
1.3 Alternative means of mass separation	24
1.4 Methods of ionization	34

Chapter 2   APPLICATIONS TO ORGANIC CHEMISTRY AND BIOCHEMISTRY

2.1 Volatility	41
2.2 Energetics of ionization	42
2.3 Determination of appearance potentials	45
2.4 Interpretation of the mass spectra of polyatomic molecules	48
2.5 The combination with chromatography	57
2.6 Some examples	65

Chapter 3   THE MASS SPECTROMETER IN INORGANIC CHEMISTRY

3.1 General considerations	70
3.2 The Knudsen cell	71
3.3 Metal chelates	71
3.4 High-temperature studies	77
3.5 Volatile inorganic compounds	80
3.6 Volatile organometallic compounds	83
3.7 Negative ions	86
3.8 Halogen compounds	89
3.9 Compounds of the noble gases	89
3.10 Spark source mass spectrography	91

Chapter 4 APPLICATION TO PHYSICAL CHEMISTRY
4.1 General considerations 95
4.2 The reactions of ions and molecules 96
4.3 The study of flames 104

Chapter 5 FURTHER APPLICATIONS
5.1 The mass spectrometer in geology 121
5.2 The mass spectrometer in space 129
5.3 The mass spectrometer in medicine 131

Chapter 6 CURRENT TRENDS IN MASS SPECTROMETRY
6.1 The mass spectrometer and the computer 137
6.2 Possible technical advances 142
6.3 Conclusion 147

Appendix Isotopic masses, abundances and ionization potentials 148

Further reading 154

Index 155

PREFACE

THIS book is an attempt to give an account of the development and applications of the modern mass spectrometer. It traces the origin of our ideas on the constitution of matter from experiments upon the discharge of electricity through gases. It describes the construction of the first mass spectrographs and their development into present day commercial instruments, and discusses the alternative methods of mass selection and their probable future. It surveys the application of the mass spectrometer to chemistry, biochemistry, geology, medicine and space. Inevitably this survey will only be superficial, but it is hoped that the reader will be persuaded to examine some at least of the specialist books which are quoted in the bibliography.

I am grateful to G.E.C.–A.E.I. for permission to publish diagrams of the ion optics of their instruments.

Birmingham   JOHN MAJER

*To my wife, Lillian,
for all her care and assistance
in the preparation of this book.*

Terms used in mass spectrometry

Appearance potential
The lowest energy which must be imparted to the parent molecule to cause it to produce that particular ion. This energy, usually stated in electron volts, may be imparted by electron impact, by photon impact, or in other ways.

Background mass spectrum
The mass spectrum observed when no sample is intentionally introduced into the mass spectrometer or spectrograph.

Base peak
The most intense peak in a mass spectrum. This term may be applied to the spectra of pure substances or mixtures.

Chemical ionization
Ionization resulting from the collision of a particle with a positively charged ion.

Double-focusing mass spectrograph
An instrument in which an ion beam, initially diverging in direction, where ions with the same mass and charge have different kinetic energies, is separated into beams according to the quotient mass/charge, these beams being focused on to a photographic plate. The instrument therefore uses both direction and velocity focusing.

Double-focusing mass spectrometer
An instrument in which an ion beam of given mass/charge is brought to a focus when the ion beam is initially diverging and contains ions of the same mass and charge with different kinetic energies. The instrument therefore uses both direction and velocity focusing. The ion beam is measured electrically.

Electron impact ionization
Ionization resulting from the collision of an electron with any particles, e.g. a molecule or atom.

Electron multiplier
A device to amplify current using acceleration of secondary electrons from an electrode emitting them to another electrode which in turn emits further secondary electrons so that the process can be repeated.

Faraday cylinder collector
A cylindrical conductor, open at one end and closed at the other, used to collect beams of ions.

Field ionization
Ionization resulting from the effect of a very strong electric field on any particle. The strong electric field may produce ionization in space or in a region very close to a metal surface.

Fragment ion
An ion produced by the cleavage of one or more bonds in a parent molecular ion.

Intensity relative to base peak
The ratio of the ion current to that of the base peak. A process of normalization is generally used with the base peak current taken as 100.

Ionization efficiency curve
A curve of ion current of a particular ion plotted against the energy of the ionizing electrons or photons.

Ionizing voltage
The voltage difference through which electrons are accelerated before they are used to bring about electron impact ionization.

Isotopic ion
An ion containing one or more atoms of a less abundant isotope.

Magnetic deflection
The deflection of an ion beam in a magnetic field.

Mass spectrograph
An instrument in which beams of ions are separated according to the quotient mass/charge, and in which the deflection and intensity of the beams are recorded photographically.

Mass spectrometer
An instrument in which beams of ions are separated according to the quotient mass/charge, and in which the intensities of the beams are measured electrically.

Mass spectrometry
The branch of science dealing with all aspects of mass spectrometers and the results obtained with these instruments.

Mass spectrum
The spectrum produced by a mass spectrometer, which shows ion current as a function of the quotient mass/charge as a series of peaks corresponding to different ions. The spectrum produced by a mass spectrograph shows a series of lines on a photographic plate.

Mattauch–Herzog geometry
An arrangement for a double-focusing mass spectrograph in which a deflection of $\pi/4\sqrt{2}$ radians in a radial electrostatic field is followed by a magnetic deflection of $\pi/2$ radians.

Metastable decomposition
The decomposition of an ion of mass m_1 into an ion of mass m_2 ($m_2 < m_1$) occurring after acceleration of the ion and before magnetic deflection of the ion.

Metastable ion peak
The peak at mass m_2^2/m_1 resulting from a metastable decomposition.

Molecular anion
An ion formed by attachment of one electron to the original molecule.

Molecular ion
The ion produced by the loss of one electron from the original molecule.

Negative ion mass spectrum
A mass spectrum produced by negative ions.

Nier–Johnson geometry
An arrangement for a double-focusing mass spectrometer in which a deflection of $\pi/2$ radians in a radial electrostatic field analyser is followed by a magnetic deflection of $\pi/3$ radians. The electrostatic analyser uses a symmetrical object-image arrangement and the magnetic analyser is used asymmetrically.

Parent ion
Ion precursor of a fragment ion or of a metastable intermediate.

Peak height
The height of a peak in a mass spectrum.

Photographic plate recording
The recording of ion beams by allowing them to strike a photographic plate which is subsequently developed.

Photoionization
Ionization resulting from a collision of a photon with any particle which is in consequence ionized.

Quadrupole mass analyser
An arrangement in which ions with a desired quotient of mass/charge are made to describe a stable path under the effect of a static and high frequency electric quadrupole field, and are then detected. Ions with a different mass/charge are separated from the ions detected because of their unstable paths.

Quistor
A three-dimensional quadrupole analyser consisting of a ring and two spheres.

Radial electrostatic field analyser
An arrangement of two conducting sheets forming part of a cylindrical capacitor giving a radial electrostatic field which is used to deflect and focus ion beams.

Rearrangement ion
An ion produced by the rearrangement of chemical bonds.

Resolution—10% valley definition
Let two peaks of equal height in a mass spectrum at masses m and $m - \Delta m$ (a.m.u.) be separated by a valley which at its lowest point is just 10% of the height of either peak. For similar peaks at a mass exceeding m, let the height of the valley at its lowest point be more than 10% of either peak height. Then the resolution (10% valley definition) is $m/\Delta m$. It is usually a function of m. $m/\Delta m$ should be given for a number of values of m. This definition implies that, for an isolated symmetrical peak, at a distance $\pm \frac{1}{2}\Delta m$ along the mass scale from the peak maximum the peak height is 5% of the maximum peak height.

Single-focusing mass spectrometer
An instrument with which an ion beam with a given value of mass/charge is brought to a focus although the initial directions of the ions diverge.

Spark source ionization
Ionization resulting from a spark between electrodes.

Thermal ionization
Ionization of particles brought about by a high temperature for example, emission of ions from an adsorbed layer on an incandescent metal surface.

Time-of-flight mass spectrometer
An arrangement using the fact that when ions of different (mass/charge) are given the same kinetic energy, they take different times to travel through a given distance in a field-free region.

CHAPTER 1
principles of mass spectrometry

1.1. *Development of the mass spectrometer*

THE conduction of electricity through gases at low pressure has been known for a long time, but it was not until the second half of the nineteenth century that it was studied quantitatively. Under a suitably high potential difference a current was found to pass between a pair of electrodes when the pressure within the glass vessel had been sufficiently reduced. At quite low pressures, 'rays' appeared to originate at the cathode and cross the vessel in straight lines. These 'cathode rays' were shown to be streams of negatively charged particles, and from their behaviour in an electric field and in a magnetic field, J. J. Thomson was able to measure their specific charge (charge/mass). The cathode ray particles are the units of negative electrical charge ($q = e = 1·6022 \times 10^{-19}$ C) which we now call electrons. In mass spectrometry, the quantity usually used is the reciprocal of the specific charge, i.e. mass divided by charge (frequently referred to loosely as the mass-to-charge ratio). For the electron, this quantity has a value of $5·686 \times 10^{-12}$ kg C^{-1}.

In 1886 Goldstein showed that if the cathode was perforated, one could see a second beam which passed through the perforated cathode and travelled onwards in a straight line to the wall of the vessel. These 'positive rays' could also be deflected by both magnetic and electric fields, but the deflection was in each case in the opposite sense to that for cathode rays, indicating that the rays were streams of positively charged particles, with mass/charge values several thousand times greater than that of the electron. These particles were identified as the residues of the molecules or atoms which had lost one or more electrons. In Goldstein's experiment electrons were emitted from the cathode and accelerated in the potential gradient close to the cathode (fig. 1.1). On their way to the walls of the vessel they encountered residual gas

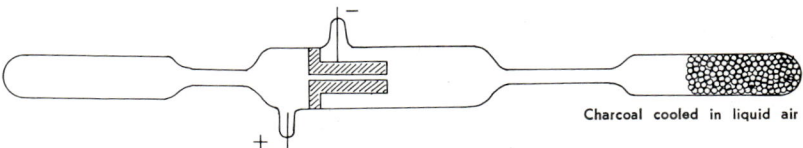

Fig. 1.1. Goldstein's apparatus.

molecules and an electron could transfer some of the kinetic energy which it had acquired to a molecule, sufficient to ionize the molecule and produce another, secondary, electron and a positive ion.

$$e + M \longrightarrow M^+ + 2e$$

The positive ions were formed almost at rest (or at least with only the thermal energy possessed by the original molecules) and were then accelerated in the potential gradient towards the cathode, through which most of them passed. The study of these beams of positive ions and the measurement of the mass/charge values for ions from different gases led to the appearance of the mass spectrometer, and incidentally to a fundamental change in our understanding of the structure of matter.

1.1.1. *The parabola mass spectrograph of J. J. Thomson*

In this instrument (fig. 1.2), first described in 1912, the beam of positive ions emerging through the cathode was passed through an analyser consisting of parallel magnetic and electric fields. These fields were made parallel and equal in length by using a pair of iron plates within the vessel which were both the electrodes and the pole faces of the magnet. The deflection of the beam in the analyser could be observed by allowing it to strike a fluorescent screen or photographic plate. If the beam of ions is considered to move in the x direction then the electric field causes a deflection in the z direction, while the magnetic field is responsible for the deflection in the y direction. The extent of these deflections is related to the intensity of these fields and to the mass-to-charge value m/q of the positive ions in the following way.

An ion of mass m and charge q moving through a homogeneous electric field V_z suffers an acceleration in the z direction of qV_z/m. If the

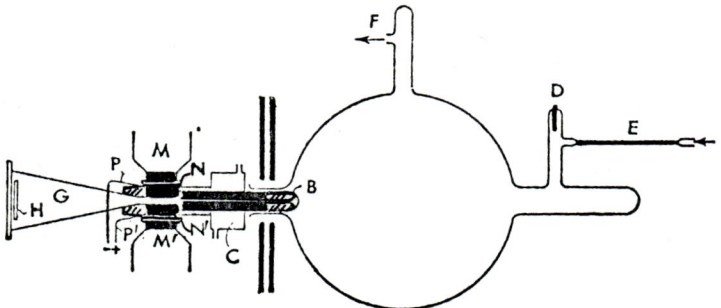

Fig. 1.2. J. J. Thomson's positive ray analyser. B, cathode: C, cooling water jacket: D, aluminium anode: E, glass capillary leak: F, pumping line: G, camera: H, fluorescent screen or photographic plate: M, M', magnet pole faces: N, N', mica insulators: P, P', electrodes.

initial velocity of the ion is v_0, then at any time t, the distance travelled along the x direction will be $v_0 t$ and the deflection suffered in the z direction during that time will be given by the expression $qV_z t^2/2m$. If t is the time taken for the ion to traverse the entire length of the electric field moving a distance x along the x axis and a distance z along the z axis, then by eliminating t

$$z = \frac{qV_z}{2m}\left(\frac{x}{v_0}\right)^2.$$

At the same time, the ion also experiences an acceleration along the y direction under the influence of the transverse magnetic field B equal to Bqv_0/m. The deflection along the y axis in time t is then given by $Bqv_0 t^2/2m$ or $Bqx^2/2mv_0$. Combination of the equations for the two deflections gives an expression of the form $y^2 = kz$ where

$$k = qB^2 x^2 / 2mV_z.$$

This equation is that of a parabola, the constant k being proportional to the specific charge q/m of the ion. k also depends upon the intensity of the fields B and V_z, the geometry of the instrument and the initial velocity of the positive ions v_0 (or the initial ion-accelerating potential V_x). Since

$$\frac{y}{z} = \frac{Bv_0}{V_z},$$

the ratio of the deflections in the y and z directions is proportional to the initial velocity of the ions v_0. As in the simple discharge tube used by Thomson the ions have a wide range of initial velocities, then all ions having the same m/q value strike the screen or photographic plate on a parabola whose vertex is the point of impact of the ions in the absence of deflecting fields. For each positive ion having a specific m/q value there is a separate parabola (fig. 1.3 illustrates this principle). If the initial accelerating potential V_x is a.c. then in alternate cycles both positive and negative ions may be accelerated and deflected, their deflections being in the opposite sense. However, the main interest is with the positive ion parabolas, and fig. 1.4 shows a photograph obtained by J. J. Thomson for the residual gases in the discharge tube. The full parabolas seen in the photograph are produced by two exposures, with the magnetic field reversed for the second exposure. The deflection in the z direction is inversely proportional to the kinetic energy of the positive ions $\frac{1}{2}mv_0^2$. This will be greatest when the ion has acquired the greatest possible acceleration from the potential gradient in the discharge tube. There will therefore be a minimum value for z which will

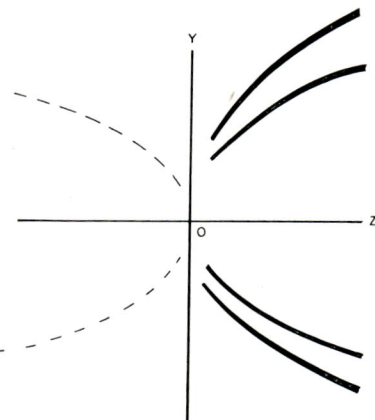

Fig. 1.3. Diagram of parabolas produced using the positive ray analyser. Vertical and horizontal lines represent the magnetic and electric axes. Continuous lines in the upper right hand quadrant represent the parabolas produced by a group of ions differing in mass. The continuous lines in the lower right hand quadrant represent parabolas which would be produced by the same group of ions on reversing the magnetic field. The dotted lines in the upper and lower left hand quadrants represent parabolas which would be produced by negative ions upon changing the polarity of the electrodes.

depend upon the potential difference across the discharge tube. The equation for the motion of the ions can be rewritten as $m/q = k_z/y^2$ so that, for any two singly charged ions of masses m_1 and m_2,

$$\frac{m_1}{m_2} = \left(\frac{y_2}{y_1}\right)^2.$$

Another way of obtaining results without the aid of a photographic plate was devised by Thomson. A parabolic slit was placed in the path of the ion beams and the ion current passing through the slit measured with an electrometer. The ion beams were presented successively to the slit by changing the magnetic field. A plot of ion current against field intensity B then showed a series of peaks corresponding to ions having a given m/q value, as seen in fig. 1.5 which constitutes the first mass spectrum.

There were two important results of this work. The first was the establishment of the mass spectrograph as a new scientific tool of great analytical power. From a single exposure it was possible to detect all of the components of a gaseous mixture while using only a very small

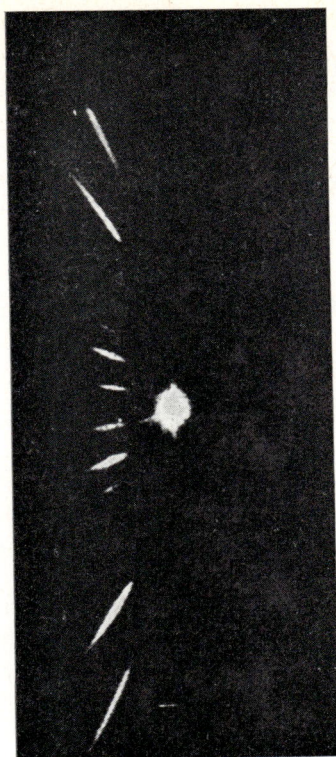

Fig. 1.4. Photographic record of positive ray parabolas from residual gases.

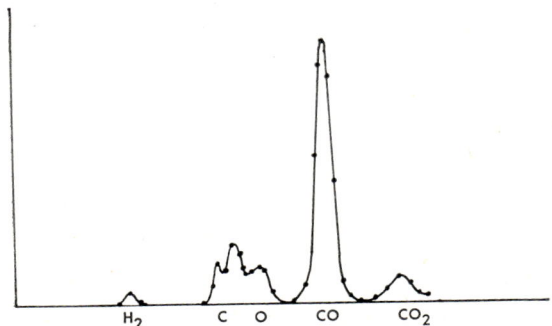

Fig. 1.5. The first mass spectrum.

mass of material. Thus in the record from the residual air in the discharge vessel it was possible to identify parabolas due to water vapour, oxygen, nitrogen, carbon dioxide and mercury (from the mercury vapour pumps used to exhaust the vessel). Other gases such as carbonyl chloride gave quite different and easily interpreted records when admitted to the instrument.

The second result was the discovery of stable isotopes. The noble gases were only just being isolated in the pure state in the period immediately before the first World War, and when Thomson examined samples of neon in the mass spectrograph he obtained not one parabola but two. The most intense was due to an ion of mass 20 times the mass of the hydrogen atom. This was close to the value expected from the vapour density of neon, which suggested a relative atomic mass of 20·2. The second and fainter parabola was due to an ion of mass 22. Although F. Soddy had shown earlier that atoms could have identical chemical properties but quite different radioactive behaviour and had introduced the term ' isotope ', this was the first demonstration of the existence of stable isotopes. This discovery was possible because the resolution obtainable with Thomson's apparatus could discriminate between masses to one part in 15. Resolution (which is discussed later) may be defined here as the ability to separate adjacent parabolas upon the photographic plate. Thomson's mass analyser gave dispersions in the y and z directions, but had no focusing properties. Improvements in resolution could only be obtained by reducing beam width. This was achieved most simply by using cathodes with a very fine central hole, but there were practical limits to the increase in resolution which could be obtained in this way.

1.1.2. *Aston's mass spectrograph*

F. W. Aston, who had collaborated with Thomson in much of the early work with the mass spectrograph, was stimulated by the need to make more precise measurements of the masses of isotopes into designing an entirely new system of mass analysis. He realized that apart from the problem of ion beam collimation, the main difficulty was that ions having the same m/q value had widely differing initial velocities, and he devised a method of overcoming this drawback of the discharge tube as an ion source. His solution was to separate the functions of the electric and magnetic fields, by separating them in space.

He used the electric field to disperse a collimated beam of ions according to *velocity*. An aperture then selected a narrow beam homogeneous in velocity, which was then dispersed according to mass in a magnetic field (fig. 1.6). The source of positive ions was still the

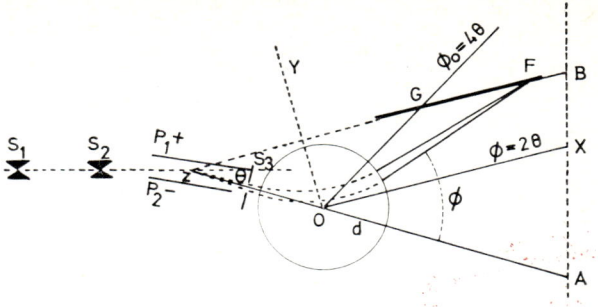

Fig. 1.6. Aston's mass spectrograph.

discharge bulb, but the hollow cathode was replaced by a pair of slits s_1 and s_2 which collimated the beam, which then passed between a pair of plates p_1 and p_2 (with a potential difference of about 500 V) which dispersed the beam according to velocity. Here, the beam may be considered to diverge from a virtual source halfway along the electric field. A further slit s_3 selected a small segment of the dispersed beam and allowed it to pass between the circular pole faces of an electromagnet. The direction of the magnetic field was chosen so as to reverse the deflection caused by the electric field. On emerging from the magnetic field the ion beam, now dispersed as to mass, fell upon a photographic plate. For positive ions of velocity v and a mass-to-charge value m/q suffering a small deflection θ in the electric field of intensity V_z,

$$\theta v^2 = l v_z q/m,$$

where l is the length of the ion path in the field. These ions then pass through the slit s_3 and the magnetic field B where they suffer a deflection ϕ and

$$\phi v = LBq/m,$$

where L is the length of the flight path in the magnetic field. For small values of θ and ϕ and constant values of V_z and B, l and L are constant. For ions with a given m/q value both θv^2 and ϕv are also constant. If $\delta\theta$ is the small change in deflection corresponding to a small change in ionic velocity δv, then for ions of a fixed m/q value but varying velocity,

$$\frac{\delta\theta}{\theta} + \frac{2\delta v}{v} = 0$$

and similarly for the deflection in the magnetic field

$$\frac{\delta\phi}{\phi}+\frac{\delta v}{v}=0$$

and

$$\frac{\delta\theta}{\theta}=2\frac{\delta\phi}{\phi}.$$

If it is assumed that the magnetic field is concentrated at the centre of the pole faces O, then let the distance $ZO = d$. For a beam of ions with a fixed m/q value, but a range of velocities, the width of the beam at O will be $d\delta\theta$ and at a further distance r beyond O the width increases to D where

$$D = d\delta\theta + r(\delta\theta + \delta\phi)$$

or $$D = \delta\theta\left[d + r\left(1 + \frac{\delta\phi}{\delta\theta}\right)\right]$$

and $$D = \delta\theta\left[d + r\left(1 + \frac{\phi}{2\theta}\right)\right].$$

However, θ and ϕ are deflections in opposite directions so θ may be considered as a negative angle and provided that ϕ is greater than 2θ, D tends to zero at a value of r given by

$$r = \frac{d \cdot 2\theta}{(\phi - 2\theta)},$$

that is to say the beam of ions becomes focused at

$$r \cos(\phi - 2\theta)/r \sin(\phi - 2\theta)$$

with reference to the axes OX and OY. The focus point is indicated at F on the line ZB, and for a constant position of the slit S_3 beams of ions of different m/q values are all brought to a focus along the line ZB which is parallel to OX. Thus a photographic plate placed along ZB records a series of focused images of the exit slit, each image corresponding to ions of a specific m/q value. It can be shown that when ϕ is approximately equal to 4θ the mass scale is nearly linear with distance along the photographic plate. This arrangement of electric and magnetic fields has energy focusing properties but is not direction focusing, hence the need for good collimation by the slits s_1 and s_2. In its original form it provided a mass resolution of 130.

Aston was able to demonstrate almost immediately that neon was indeed composed of two species of mass 20 and 22, and subsequent exposures revealed the existence of a third isotope of mass 21. In this way a new era began, in which it was demonstrated that the multiplicity of stable isotopes observed in neon was the rule rather than the exception. In succeeding years more and more isotopes were discovered and at the present time a very large number are known. Many of these are listed in the Appendix. A further advance came, however, when Aston began to examine the exact masses of the elements. In a series of the most painstaking experiments, in which the accuracy of his mass measurements was improved by taking the mean of numbers of determinations, Aston was able to demonstrate small divergences from the whole number rule. In particular, with an accuracy of 1 part in 1000 he found mean values of 6·008 for ^{6}Li and 55·95 for ^{56}Fe. He identified the significance of these discrepancies in his book published in 1924. He realized that the approach of two charged particles to distances comparable with the size of the nucleus would require the expenditure of a large amount of energy and this would be reflected in a reduction in mass. He replaced the whole number rule by the statement that the mean packing effect in all atoms is approximately constant. By 1933 Aston had improved the accuracy of his determinations by an order of magnitude and had developed the concept of packing fraction. This was defined as the percentage difference between the isotopic mass of an atom and its mass number. It represented the divergence of the atom from the whole number rule, divided by its mass number. The variation of this divergence with atomic number was illustrated most effectively in the famous packing fraction curve reproduced in fig. 1.7.

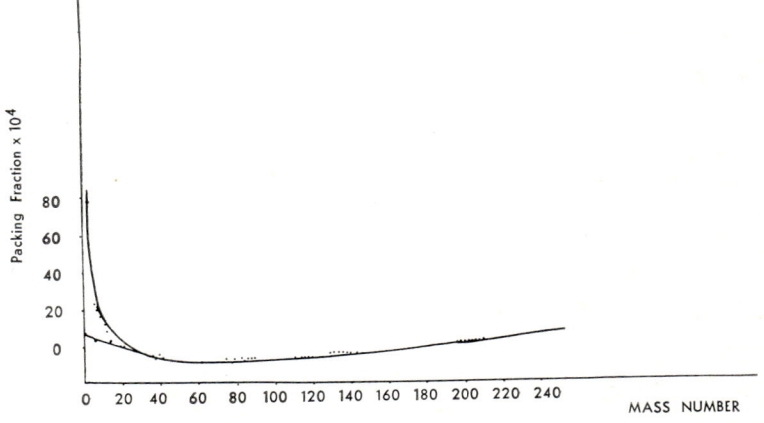

Fig. 1.7. Aston's packing fraction curve.

This curve helped to explain many nuclear phenomena, including the release of energy on the radioactive decay of the heavy elements and the stability of the light nuclei to alpha particle bombardment. The implications of the mass defects in nuclei were readily appreciated by Aston, and he gave a very clear forecast of the age of atomic energy. His calculation of the energy which would be released by the transmutation of one gram atom of hydrogen into helium has been quoted ever since.

1.1.3. *Dempster's 180° focusing mass spectrometer*

The great disadvantage of the gas discharge bulb as an ion source was that the ionization took place in the cathode space. Here electrons ejected from the cathode with a range of velocities could collide with gas molecules over considerable distances from the cathode face, so that ions did not all receive the same acceleration and acquire the same final velocity v_0 before passing through the cathode and reaching the analyser region. Dempster realized that if all the ions of a given m/q value could be given the same velocity, then a transverse magnetic field alone would be sufficient to carry out mass analysis. His solution to the problem was to separate the ionization and acceleration processes.

Ions were formed by collision with electrons obtained by thermionic emission from a hot tungsten filament and accelerated in a voltage gradient which was small compared with that used for the acceleration of the ions. All the ions were formed in a comparatively small volume and given the same acceleration from a much larger potential gradient. The ions were then subjected to the action of a magnetic field at right angles to their line of flight. As a result the ions suffered a torque in a direction perpendicular to both the line of flight and the direction of the field. Thus in a homogeneous field the ions described circular paths.

Dempster, working in Chicago, in 1918 published an account of the first direction focusing mass spectrometer (fig. 1.8). In this design positive ions from an ion source, accelerated by passing through a potential gradient V_x, enter a uniform magnetic field B which is at right angles to their line of flight. The ions with mass m acquire a velocity v_0, and hence have a kinetic energy $\frac{1}{2}mv_0^2$ which is equal to the energy given to the ion by moving its charge q through the potential difference V_x;

$$\tfrac{1}{2}mv_0^2 = qV_x.$$

As a result of the interaction of the moving charge and the transverse magnetic field B, the ion suffers a torque Bqv_0 causing it to move in an arc of radius R. The ion is constrained to move in its circular orbit by the balancing centripetal force which is equal to mv_0^2/R, thus

$$mv_0^2/R = Bqv_0.$$

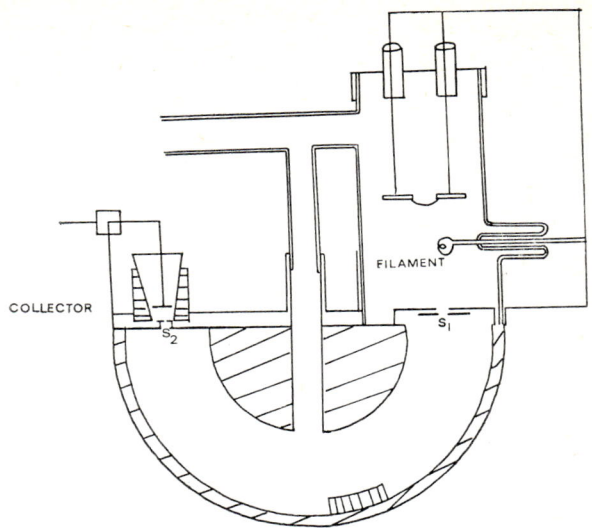

Fig. 1.8. Dempster's mass spectrometer. S_1, ion source exit slit: S_2, collector entrance slit.

So, eliminating the velocity v_0 between these two equations, the magnetic field and potential gradient requirements for an ion of mass m and charge q moving in a circular orbit of radius R are given by:

$$\frac{m}{q} = \frac{B^2 R^2}{2V_x}$$

In S.I. units, m must be in kg, q in coulombs, B in tesla and V_x in volts.

The uniform magnetic field has a focusing action which is equivalent to that of a cylindrical lens. Beams of ions issuing from a slit s_1 are brought to a focus after suffering a deflection of 180° in a circular path. An image of the source slit s_1 is formed at the collector slit s_2 for ions of the m/q value appropriate to the accelerating potential V_x and field strength B. This system is a direction-focusing arrangement, so that it is not necessary to inject a parallel beam of ions into the magnetic field. On the other hand, it requires that all ions having the same m/q value should have the same velocity, so that it is not an energy or velocity focusing system. Thus the high voltage a.c. discharge applied to the early positive ray analysers was quite inappropriate. A simple method of producing ions was the thermal ionization source in which the sample was painted directly on the filament and evaporated in the form of positive ions. In Dempster's design these ions, after passing through the slit s_1, travelled in an orbit and passed through the collector slit s_2, where they encountered the collector electrode. The current collected by this electrode was measured with a sensitive electrometer. The m/q value of the ions passing through the slit s_2 could be selected by altering the ion accelerating potential V_x, as the magnetic field was provided by a large pole face permanent magnet and remained invariant. Thus scanning was achieved by continuously altering V_x. From the equation it will be realized that the mass scale was linear and easy to calibrate. The resolution of this type of instrument depends on the stability of V_x, on the homogeneity of the magnetic field over the whole of the ion flight path, and also on the width of the source and collector slits s_1 and s_2. The optimum values for slit widths are those in which a successful compromise has been made between the increase in resolution obtained by reducing the slit widths and the accompanying reduction in sensitivity.

The original definition of mass resolution used by Dempster is expressed in the equation:

$$m/\Delta m = R/(s_1 + s_2),$$

where m is the mass of the ion and Δm is the mass increment which can be resolved, s_1 and s_2 are the slit widths as before and R is the radius of the flight path. This assumes that the image is perfect and that when one image of the source slit occupies the aperture of the collector slit none of the ion current from the adjacent image enters the same aperture. This is not true because the ion optical system is not perfect. The image is distorted, suffering from chromatic aberration because of the narrow distribution of velocities for ions having the same m/q value. It also has subsidiary side peaks due to scattering in the flight path as a result of ion–molecule collisions. These aberrations are proportional to the radius of the orbit R and the proportionality constant β so that:

$$m/\Delta m = R/(s_1 + s_2 + \beta R).$$

This gives the theoretical resolution, but it is more common to give an empirical definition by measuring the mass difference between adjacent peaks of equal height for a given percentage of the peak height above baseline in the valley between the two peaks. The value of the acceptable peak height percentage varies from 1 to 50 according to the whim of the manufacturer, so that it is well to check this criterion before comparing the performance of competitive instruments.

The effect of slit width upon the attainable resolution using a modern 180° focusing design is illustrated in fig. 1.9, which shows the spectrum of the krypton isotopes with slit widths of 0·25 and 0·50 mm. The modest resolution obtainable with Dempster's original machines was

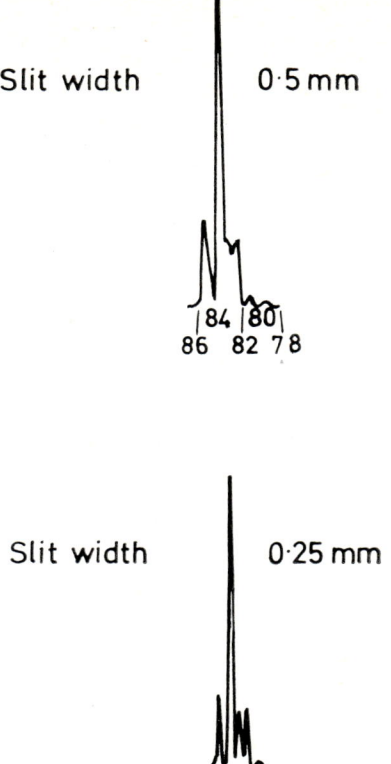

Fig. 1.9. Variation of resolution with slit width. Mass spectrum of the isotopes of krypton.

sufficient to allow the isotopic constitution of a large number of elements to be measured. The electrical detection system made it possible to measure isotopic abundance with sufficient accuracy to detect variations in elemental isotopic constitution with place of origin. This established the mass spectrometer as a tool in the study of geology. Later, during the 1930's, Bainbridge combined the 180° deflection system with a filter to reduce the velocity spread and increase resolution, and Bleakney replaced the permanent magnet with an electromagnet. It is interesting that Dempster's original concept has survived and is incorporated in some present day low resolution mass spectrometers.

1.2. Commercial mass spectrometers

The MS10, manufactured by the A.E.I. Company, originally used a large permanent magnet with a field strength B of 0·185 T and a flight path radius R of 5 cm. The resolution which could be achieved with a slit width of 0·5 mm was 40, with a 2 per cent valley between adjacent peaks. The mass scale, which was linear, could be scanned between 1 and 200 by adjusting the ion accelerating potential V_x continuously or in steps. The resolution has been steadily increased by decreasing the slit widths and increasing the sensitivity of the ion detection system, and this type of instrument is still widely used in measurements of isotopic abundance. More recently, while retaining the original dimensions, it has been completely redesigned by providing a large pole-face electromagnet which allows the spectrum to be scanned magnetically. The slit width has been reduced to 0·025 mm and the resolution thereby increased to 600. The resulting loss in ion current has been made up by using as an ion detector the more sensitive electron multiplier. This device may be described briefly as a stack of plates carrying increasingly positive voltages. The ion strikes the first plate, and electrons are ejected. These are accelerated to the second plate in the potential gradient and, on striking it, eject a further shower of electrons. This process is repeated up to 17 times, so that there may be a million charges leaving the final plate for one arriving at the first plate. This type of detector is used on all modern high sensitivity machines. The 180° deflection instrument with these facilities was given the model name of MS20, and it incorporated a direct connection to a gas chromatograph. Further, the rate of scan was sufficiently high to permit the mass spectrum to be displayed upon an oscilloscope, as well as being recorded upon a fast response recorder. Fig. 1.10 shows such a machine in operation.

In 1940 Nier published in the United States an account of an entirely new design of mass analyser, the direction focusing magnetic sector instrument (fig. 1.11). He showed that the deflection through 180° was just a special case of deflection through any angle. The criterion for successful focusing was that the ion beam would enter and leave the magnetic field with the field boundaries at right angles to its line of flight. In order to reduce magnet size he chose an angle of 60° for his sector field and had both the ion source and collector outside the magnetic field, symmetrically disposed with respect to the magnet. The advantage of this arrangement was that both collector and source were accessible without disturbing the magnet, but the long flight path made alignment critical and increased the chance of scattering by residual gas molecules. Nevertheless, this was the prototype for a very large

Fig. 1.10. The A.E.I. MS20 Gas Chromatograph/Mass Spectrometer.

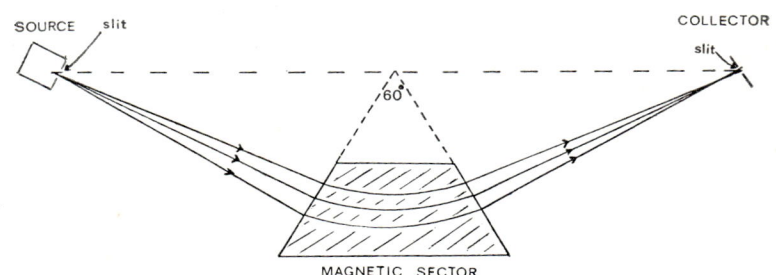

Fig. 1.11. Ion optics of a 60° magnetic sector designed by Nier.

number of spectrometers, and the design was not superseded until the advent of the large double-focusing machines described later. Another feature for which Nier was largely responsible was the electron bombardment ion source in which the sample was subjected to a constant controlled beam of low voltage electrons.

Later, Hipple produced a 90° sector instrument and, since then, all magnetic sectors in single-focusing machines have had angles of either 60° or 90°. The new magnetic sector instruments had mass resolutions sufficient to separate the isotopes of uranium, and so they played an important role in the nuclear fission programme of the early war years. The first few traces of pure ^{235}U were separated using a mass spectrometer, and later scaled-up versions called calutrons were used in routine electromagnetic separations of isotopes. The ability of magnetic sector instruments to detect and identify traces of organic compounds, and the fact that geologists were now familiar with the idea of mass spectrometry as an aid in prospecting, stimulated the commercial production of an instrument for detecting traces of petroleum in soil. In 1943 the first commercial mass spectrometer was sold by the Consolidated Electrodynamic Corporation in Pasadena to the Atlanta Refining Company. Later it was used to examine the mixtures of hydrocarbons obtained in petroleum distillates. When the War ended, the demand for commercial mass spectrometers increased and the instrument manufacturers could foresee a stable market for their products. The atomic energy programme required instruments capable of following the enrichment of ^{235}U, and the rise of the petrochemical and synthetic polymer industry created a demand for a mass spectrometer capable of analysing a wide range of organic compounds.

In the U.K., the Metropolitan Vickers Company (later G.E.C.–A.E.I.), produced a magnetic sector instrument in 1951. This was one of the most successful designs to be marketed; it was sold all over the world and many of the early models are still being operated. The MS2 was a 90° magnetic sector instrument with a radius R of 15 cm. It used an ion accelerating voltage V of 2 kV and an electromagnet providing a field B which could be varied from 0·05 to 0·6 T. The spectrum could be scanned magnetically between masses 1 and 250, and higher masses could be reached by reducing V. The mass resolution was 200 and the spectrum was recorded by an ingenious potentiometer pen recorder, which automatically changed range with the strength of the signal to be recorded. This device, now obsolete, allowed mass peaks with a ratio in height of 10^4 to be drawn on the same chart paper. By modern standards the instrument was slow, requiring 20 minutes for a complete scan, but it had some very interesting general facilities. The electron accelerating potential could be adjusted between 5 and 100 V and could be measured accurately, so that ionization and appearance potentials could be determined from plots of ion current against electron voltage.

The collector system was designed to permit the simultaneous detection of two adjacent mass peaks at different sensitivities. This was achieved by collecting one ion current in the normal way and a second ion current at an adjacent m/q value on the plate carrying the defining slit through which the first ion current passed to the normal detector. The first ion current due to the less abundant isotope was small and was amplified at high gain, and the resulting signal fed to the tapping point of a potentiometer. The second ion current, due to the more abundant isotope, was amplified at low gain and the resulting signal fed to the ends of the potentiometer. The two signals which were related to the two isotope abundancies could therefore be compared simultaneously, so avoiding errors due to changes in instrumental sensitivity. This device made it possible to determine small variations in the abundance of rare isotopes. As an example, when studying the incorporation of ^{15}N into organic substances, the labelled compounds were converted to nitrogen gas and the ^{15}N abundance determined from the ratio of the ion currents at m/q 28 and 29. The main function of the instrument was, however, the analysis of organic compounds of low molecular mass. The success of the instrument stimulated some further developments. The slit widths were reduced and an electron multiplier provided, thus improving the mass resolution to 1000. There was also a steady improvement in sample introduction systems. Originally the organic compound had been admitted at low pressure through a porous ceramic disc at room temperature, a method which restricted samples to those boiling below 200°. Later, higher-boiling samples were accommodated by injecting them into a heated reservoir through porous discs protected from atmosphere by pools of molten gallium. There were many variations and improvements upon Nier's magnetic sector, but the design finally became obsolete with the arrival of the double focusing system.

1.2.1. *Double-focusing instruments*

During the 1950's the resolution of the magnetic sector instrument was gradually improved, mainly by decreasing the source and collector slit widths and increasing the sensitivity to compensate for the loss in ion current. From the simple equation for the calculation of resolution ($m/\Delta m$), it would seem that this improvement could be continued indefinitely. This is not the case, however, because the beam of ions produced by a conventional ion source is not sufficiently homogeneous either in velocity or direction. The simple equation for the path of ions in a magnetic sector would only be true for all ions in a beam if they all had the same direction, and if all ions with the same value of mass-to-charge had exactly the same velocity. Any divergence from these requirements results in such ions describing different paths and arriving at a different point at the collector. As a result, the ion beam for any particular species is broader and more diffuse at the collector than it is at the source. Consequently resolution is reduced below the theoretical value. The most serious of these factors is the velocity spread. In theory all ions having the same mass-to-charge value should receive the same acceleration in the high potential gradient and acquire the same terminal velocity. However, not all ions are produced in exactly the same plane and they have a thermal distribution of velocities. Analysis

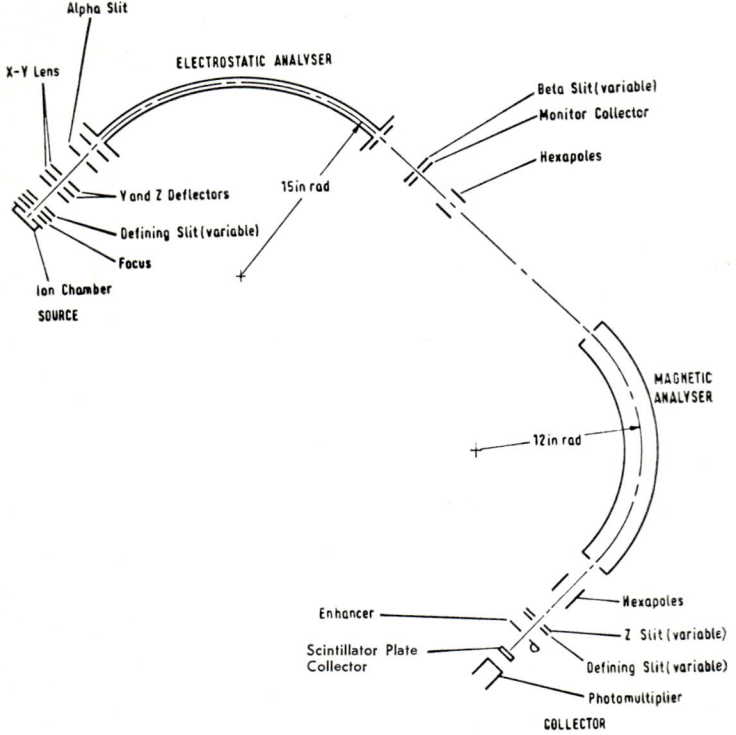

Fig. 1.12. Ion optics of the double-focusing system (MS50 Series). Nier–Johnson geometry.

Fig. 1.13. The A.E.I. MS9 Double-Focusing mass spectrometer.

of beams of similar particles having different velocities had been achieved many years previously by passage through a transverse electrostatic field, i.e. between the plates of a condenser, but it was Mattauch and Herzog who first defined the conditions for focusing of beams of ions for both direction and velocity. They formulated the principle of double focusing using two separate stages. They suggested the combination of a conventional magnetic sector with a second electrostatic sector in the form of a radial electrostatic field, for which:

$$V_z \frac{dV_z}{dR'} = \text{constant},$$

where R' is the radius of the flight path of the ion and V_z is the potential. These conditions could be realized by using a pair of curved plates or hemispheres between which the ions passed. The voltages on the plates were related to the ion accelerating potential by the equation

$$R' = \frac{2V_x}{V_z}.$$

The first successful commercial double-focusing mass spectrometer was described by Craig and Errock. It was based upon the double-

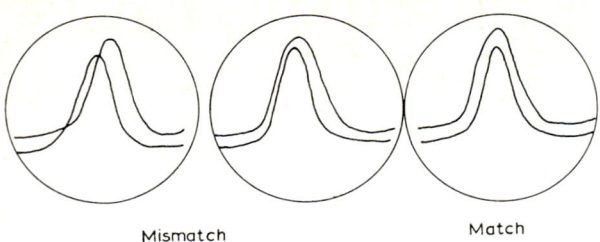

Mismatch   Match

Fig. 1.16. Determination of precise mass by peak matching.

peaks could appear on the screen alternately (fig. 1.16). The ion-accelerating voltage for one scan was the full voltage, while that for the second scan was a fraction of that voltage obtained with the aid of a very accurate six-decade resistance box. By altering the decades on the box, one peak could be made to move across the screen relative to the other and a point was normally chosen where the two peaks were superimposed. The ratio of the masses of the ions providing the two peaks was then equal to the ratio of the two accelerating voltages, as indicated by the dials of the decade box, $m_1/m_2 = V_2/V_1$. It was thus only necessary to know the mass of one ion to calculate the precise mass of the other. In practice, a calibrating substance was used, whose spectrum and ionic masses were known to the operator and the peaks of the unknown sample were compared with those of the known sample. As the double focusing system could provide mass resolutions of the order of 50 000 and it is possible to determine relative peak positions with an accuracy of 10 times the resolution, the two masses could be compared with an accuracy of a few parts per million. Recently, this figure has been improved to one part in 10^7.

The ability to make mass comparisons with this order of accuracy enabled unknown ions to be identified unequivocally by their mass alone. The present mass scale is based upon the element carbon, the ^{12}C isotope being given a mass of 12 exactly. Upon this scale hydrogen H has a mass of 1·007825, while oxygen ^{16}O and nitrogen ^{14}N have masses of 15·994914 and 14·003074 respectively.

If we consider a simple example of two ions with the same integral mass, the molecule ions of carbon monoxide and nitrogen with a mass of 28, we find that the accurate masses, 27·994914 and 28·006148 respectively, are appreciably different. It is thus possible to distinguish between and identify these two closely similar ions. The precise masses of all common isotopes are listed in the Appendix, and it is possible to calculate the precise mass of any polyatomic ion by simple addition. In practice, the experimentally determined precise mass of an ion is compared with values for the precise masses of all polyatomic ions listed in Tables, and the atomic constitution of the unknown ion is thereby identified. These features often allow the chemist to record a mass spectrum using only a few micrograms of an involatile sample, and find not only the atomic constitution of the molecule (provided that the molecule ion was present) but also the atomic constitutions of fragment ions, thus providing some structural information. This feature has also now been overtaken by the greater flexibility of the mass spectrometer–computer combination. At present the most advanced instrument is the MS50 which largely retains the geometry of the MS9 but has all of its facilities transistorized and its output computerized. In addition, however, it can be used to study the so-called metastable peaks which are due to ionic dissociations taking place outside the ion source. By selecting voltages on the electrostatic analyser it is possible to eliminate the normal mass spectrum and record only those peaks due to ions resulting from such dissociations. This can provide information about the mechanism of ionic dissociation, and also estimates of the kinetic energy released in such dissociations. This has led to the development of a separate sub-discipline called ion kinetic energy spectroscopy, or I.K.S.

1.2.2. *Double-beam instruments*

Probably the most interesting development in instrument design over the past decade has been the invention of the double-beam mass spectrometer. Most optical spectrometers are double-beam instruments, this facility being used to cancel out the effect of instrumental blanks, cell absorption and solvent effects. It was only recently, however, that the problem of constructing a double-beam mass spectrometer was overcome. The principle involves the passage of two ion beams through the same mass analyser systems, so that two separate vacuum systems, ion sources and collectors must be used, but only a single magnet.

In the MS30, marketed by the A.E.I. Company, this is achieved by passing the two ion beams through modified electrostatic sectors so that they become parallel. After passage through the magnetic field they are once again deflected apart to two separate detectors (fig. 1.17). The output of these two detectors can be fed to a two-channel logarithmic-response galvanometer recorder, to give two traces side-by-side. The general advantages of this design are obvious. The spectrum of an unknown compound may be compared directly with that of an authentic sample or of a mass marking substance. Instrumental blank peaks

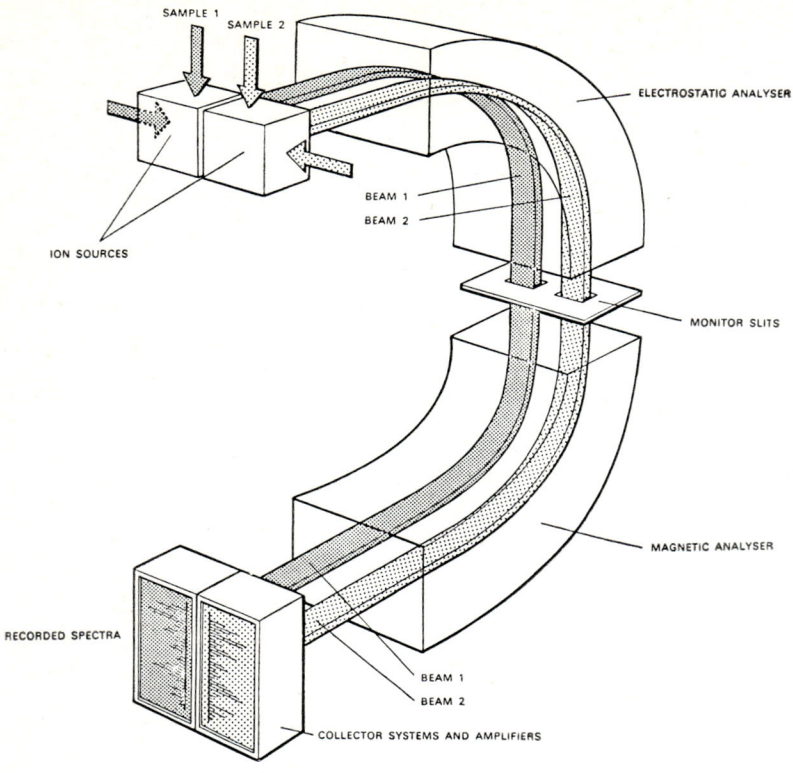

Fig. 1.17. The ion optics of a double-beam mass spectrometer.

are easily identified and eliminated. The same substances may also be examined simultaneously under different conditions. Thus the effect of temperature, pressure, electron voltage and ion-accelerating voltage may readily be assessed. Further, since a double focusing system is being used, all the facilities described in the previous section are available. In addition to this essentially high resolution instrument, smaller, less expensive, low resolution designs are being produced.

1.3. *Alternative methods of mass separation*

There are probably over two dozen methods of segregating ions according to their values of mass divided by charge, but only a few have received commercial exploitation. Instruments based upon the determination of ionic velocities and ionic motion in radio-frequency fields are now discussed.

1.3.1. *The time-of-flight principle*

The concept of measuring the mass of a particle by determining the velocity it acquires under the influence of a known potential gradient dates from the end of the nineteenth century. Weichert applied the idea to the measurement of the m/q value for cathode rays. The extension of the principle of the velocity filter to the measurement of the m/q values for comparatively heavy ions has given rise to the construction of many varieties of what have been called time-of-flight mass spectrometers. Almost all designs consist of a vacuum system involving a long, straight tube down which the ions drift under the influence of an accelerating potential gradient. At one end of the tube is situated the ion source and at the other end is the collector. Because of the very high velocity which ions acquire in even the most modest potential gradients, drift tubes of the order of one metre in length require time resolutions of the order of a microsecond. The principle is to produce a bunch of ions by operating either the electron-accelerating or the ion-accelerating potentials in a pulsed mode. When the bunch of ions has been formed, all the ions theoretically experience the same force from the ion-accelerating potential, and therefore acquire a velocity which is related to the square root of their mass. The ions drift along the tube and strike the collector. This collector may be operated in a pulsed mode, accepting only those ions which have taken a predetermined time to pass down the tube, or it may be continuously active. If the output of the collector is displayed upon a cathode ray oscilloscope, whose sweep rate is synchronized with the repetition rate of the source pulses, the trace on the screen constitutes a mass spectrum. The peak heights correspond to the ion abundances, and the position along the x axis is related to the time of arrival at the collector and hence to mass. The time-of-flight t for ions of mass m and charge q acquiring a constant energy qV_x from an accelerating potential V_x is given by

$$t = L\,(m/2qV_x)^{1/2}$$

where L is the length of the drift tube.

This simplified description takes no account of the problems involved in the design of either the ion source or the collector, let alone the pulsing circuits, for even a modest resolution of 200 requires a pulse width of less than $0.1\ \mu s$ for a flight tube one metre long. With bursts of ion production lasting only 0.01 s it was possible to consider repetition rates as high as $10\,000\ s^{-1}$. This feature, combined with oscilloscopic display, led many potential mass spectrometrists to envisage a machine which would give records of the variation of sample composition lasting over only a tenth of a millisecond, as in, for example, fast thermal or photochemical reactions. Such hopes, however, have been only partially realized, because of the difficulty of making enough ions in 0·01 s to provide a statistically reliable mass spectrum. This has led in turn to the custom of integrating the signals received over a hundred cycles by the use of a collector with a gate, which accepts only those ions with a specific flight time and hence mass range.

The resolution of the time-of-flight instrument is limited largely by the fact that ions are produced over a small but finite volume with an initial kinetic energy which is related to the temperature of the gas molecules. This means that not all ions will be formed at the same point and therefore they have different flight paths. In addition, the initial thermal velocity may be in the same direction as the final ion velocity or in the opposite direction. Refinements in the source design have increased the resolution until it approaches 1000, but it seems fairly certain that really high mass resolution will not be achieved with this design. While the sensitivity is theoretically higher in the time-of-flight design, because of the absence of restricting slits, the necessity of rejecting signals which are harmonics of the fundamental mass signal reduces transmission and leaves the effective sensitivity about the same as for magnetic sector instruments. However, the time-of-flight mass spectrometer retains the advantages of rapid mass scan and simple open source structure, and for this reason it has been chosen in studies of ion–molecule reactions. The most successful commercial instrument is that produced by the Bendix Corporation of America.

1.3.2. *Cyclotron resonance mass spectrometers*

The cyclotron was one of the early particle-accelerating machines used in studies of nuclear physics. It can be thought of as a mass spectrometer in which the ions travel in circular paths under the influence of a transverse magnetic field. In the original machine the ions, usually protons, were accelerated by a radio-frequency field and this frequency, which is called the cyclotron resonance frequency μ, is given by the equation:

$$\mu = Bq/2\pi m,$$

where m is the mass of the proton and μ is derived from the magnetic deflection equation quoted earlier, $mv_0^2/R = Bqv_0$, by substituting $v_0 = 2\pi R\mu$. If the cyclotron can be adapted to deal with ions of higher mass, it becomes possible to determine ionic mass by the measurement of the resonant frequency. This is an attractive idea since the measurement of frequency is one of the more accurate physical measurements. The most elaborate of these instruments was the mass synchrometer built by L. G. Smith at Brookhaven. Here an ion beam took up a helical path in the field of a powerful electromagnet. It left the ion source and was directed into its path by passage through a slit which was connected to an oscillator operating at the resonant frequency. It passed through additional slits after three revolutions and was deflected to a detector by the final slit, which was also connected to the oscillator. Frequencies of about 15 MHz were used for path diameters of about 25 cm. The radio-frequency may be swept so as to display the output of the detector on an oscilloscope in the form of a peak. Very precise mass comparisons were made possible by carrying out alternate sweeps with different accelerating voltages and adjusting these voltages until the peaks became coincident.

1.3.3. R.f. field techniques

(a) The mass filter

In the early nineteen fifties a revolutionary new form of mass analyser was developed based upon the oscillations of ions passing through an axially symmetrical radio-frequency field. We have already discussed the deflection of beams of ions in an electrostatic field which is invariant with time. Paul and von Zahn considered what would happen if the field varied very rapidly, so that an ion could suffer only a small deflection before the direction of the torque it experienced was reversed with a consequent reversal in the sense of its deflection. The extent of the oscillation of the ion traversing the field will be related to either the velocity or the mass of the ion.

Paul considered the special case of an ion travelling through a radio-frequency quadrupole field, that is a field imposed by four electrodes disposed symmetrically around the flight path of the ion, i.e., the x direction. The ion will then experience deflections in y and z directions at right angles to its flight path. The ion entering the quadrupole field experiences a varying potential P of the form:

$$P = P_0(\alpha y^2 + \beta z^2 + \gamma x^2)$$

where $\alpha + \beta + \gamma = 0$, $\alpha = -\beta = 1/r_0^2$, $\gamma = 0$, and $P_0 = U + V \cos \omega t$, r_0 is the field radius, $\omega =$ frequency, $t =$ time and U and V are voltages. Therefore

$$P = (U + V \cos \omega t)(y^2 - z^2)/r_0^2.$$

The system is now defined more practically by assuming the ion to be moving along the x direction around which are symmetrically spaced four rods. The rods are connected in opposite pairs, the lines joining opposite pairs being the y and z directions. If the potentials on the rods are P_y and P_z then $P_y = -P_z = U + V \cos \omega t$. The equations of motion of the ion are then given by:

$$\frac{d^2y}{dt^2} = -2q(U + V \cos \omega t) \frac{y}{r_0^2},$$

$$\frac{d^2z}{dt^2} = 2q(U + V \cos \omega t) \frac{z}{r_0^2},$$

$$\frac{d^2x}{dt^2} = 0.$$

These are the Mathieu equations, which equate the force experienced by the ion, from the interaction of its charge and the instantaneous potential, to its mass multiplied by its acceleration in the appropriate direction. These differential equations must be integrated in order to obtain the path of the ion through the radio-frequency field and the solution is complex, but the result may be described qualitatively. The oscillatory flight path of the ion may either be a stable trajectory, which allows the ion to traverse the entire field, or be an unstable trajectory, y and z increasing with time and the amplitude of the oscillations becoming so large that the ions strike the electrodes and are lost. The conditions for the execution of a stable trajectory involve the mass-to-charge value of the ion and the U/V ratio, that is to say the ratio of the d.c. voltage U applied to the rods and the peak amplitude of the radio-frequency voltage V (i.e. when $\cos \omega t = 1$). This ratio may be selected so that only a very restricted range of m/q values satisfies the condition for stable trajectories. That is, only ions having a restricted range of masses can pass along the axis of the rods and leave the field. Thus the radio-frequency field is acting as a mass filter.

The quadrupole electrode assembly can be incorporated into a mass spectrometer by placing an ion source at the field entrance and an ion detector at the field exit. In designing such an instrument certain criteria have to be applied. The values of the radio-frequency peak amplitude V and the d.c. voltage U are related to the ionic mass, the frequency ω and the field radius r_0 by the equations:

$$V = 7 \cdot 219 \ M\omega^2 r_0^2$$

and

$$U = 1 \cdot 212 \ M\omega^2 r_0^2,$$

where r_0 is in cm, ω is the frequency in MHz and M is the relative mass of the ion. The U/V ratio is then $0 \cdot 1678$ and the radio-frequency power required is $6 \cdot 5 \times 10^{-4} \ CM^2\omega^5 r_0^4/Q$, where C is the capacity of the rod assembly in picofarads and Q is the figure of merit of the power circuit, a quantity which describes the efficiency of power conversion in the circuit. The maximum allowable injection voltage for the ions is given by $4{\cdot}2 \times 10^2 \omega^2 L^2 M \times$ (Resolution)$^{-1}$, where L is the length of the rod in metres. Thus the injection voltage is restricted to below 200 V and is sometimes less than this, providing an interesting comparison with sector instruments which normally operate in the kilovolt region. The limit to the mass range is set by the radio-frequency power requirements, and the resolution $m/\Delta m$ depends on the stability of the peak radio-frequency voltage V and the constancy of the U/V ratio.

The most uncompromising feature of a mass filter is the critical nature of the field radius r_0, which must be kept to within a ten-thousandth of its design value. The precise solution of the Mathieu equations requires the field boundaries to be hyperbolic in cross-section, but electrodes of this geometry would be difficult to machine so an approximation in which rods of circular cross-section are used has to be accepted (fig. 1.18). Using a rod-radius to field-radius ratio (r/r_0) of 1·16 is the best compromise, reducing the resolution obtainable by only a factor of 2. The radio-frequency employed must be as low as possible, consistent with the ions performing a sufficient number of oscillations within the field, as the power required goes up as the fifth power of the frequency. As an example: for a frequency of 1·59 MHz, rod length of 1 metre, capacitance of 400 pF and an ion of mass 10, the power

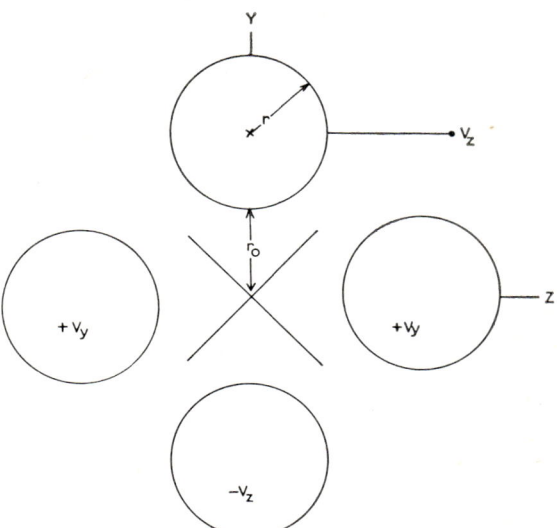

Fig. 1.18. The mass filter. Geometry for a quadrupole radio-frequency field.

required is 80 watts. In practice instruments having a wide mass range usually scan by a combination of frequency and voltage adjustments.

The advantages of the mass filter are its compact nature and absence of large, heavy magnets. In addition, it is insensitive to variations in ion energy, so that crude ion sources may be used and naturally occurring ions in the upper atmosphere can be studied. Its main disadvantage is the need to maintain a highly stable geometry for the rods to within a few μm.

(b) *Development of the quadrupole mass filter*

A number of companies in Britain, Germany and the United States have developed commercial instruments based on the quadrupole system of mass analysis. Because it is comparatively easy to construct devices with resolutions of 80 or less, the majority of these have been low resolution residual gas analysers, such as the Q4 marketed by VG-Quadrupoles Ltd.

Increases in resolution require decreasing tolerances in the rod dimensions and assembly and this sets a practical limit to resolution. Fite has calculated that to achieve a resolution of 5000 an accuracy of 1 part in 40 million is required. At the time of writing, most commercial instruments have not approached this figure. The second problem encountered by the designers was the provision of adequate power supplies for the rods. The radio-frequency power needed rises steeply with rod dimensions, increasing mass and frequency. The frequency should, therefore, be as low as possible and the rods as small as possible, consistent with the ions describing sufficient oscillations during their flight path. Frequencies chosen are often in the 1–5 MHz region, and rods about 20 cm long with a diameter of 0·6 cm are commonly used.

Fig. 1.19 is a photograph of an experimental rod assembly and fig. 1.20 shows an experimental quadrupole built in glass.

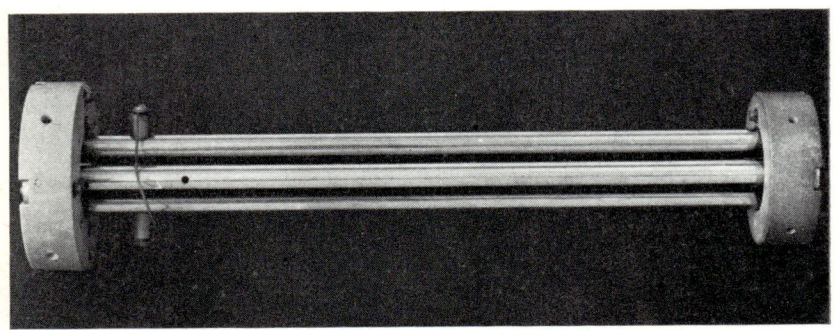

Fig. 1.19. Experimental rod assembly for a mass filter.

Fig. 1.20. An experimental mass filter used for chemical ionization.

A severe limitation of the mass filter is the fringe field of the rods experienced by the ion as it approaches the gap between the rods. The effective aperture is not the actual physical gap between the rods, but an electric aperture imposed by the fringe field. In the original commercial design produced by the Atlas Company of Germany, the ions were injected into the gap through an earthed tube. In more modern instruments the problem has been overcome by providing a second short quadrupole assembly, which carries only the radio-frequency voltage and focuses the ions into the main quadrupole field. One modern instrument marketed by Finnigan Instruments Ltd has a high performance with a mass range of 1000 and a resolution of 2000. Its rapid rate of scan allows continuous display of the spectrum and also the monitoring of a number of masses continuously, the rod voltages changing in steps rather than linearly.

This instrument is eminently suitable for direct connection to a gas chromatograph, a combination which has been successful in the study of air pollution. In addition to these general purpose instruments, specific quadrupole assemblies have been designed for tasks as diverse as the monitoring of the upper atmosphere and the study of collisions in molecular beams. In both environments its compact size and simplicity recommend its use.

There are some intriguing variations on the quadrupole geometry, the simplest being the monopole. Von Zahn was the first to suggest that

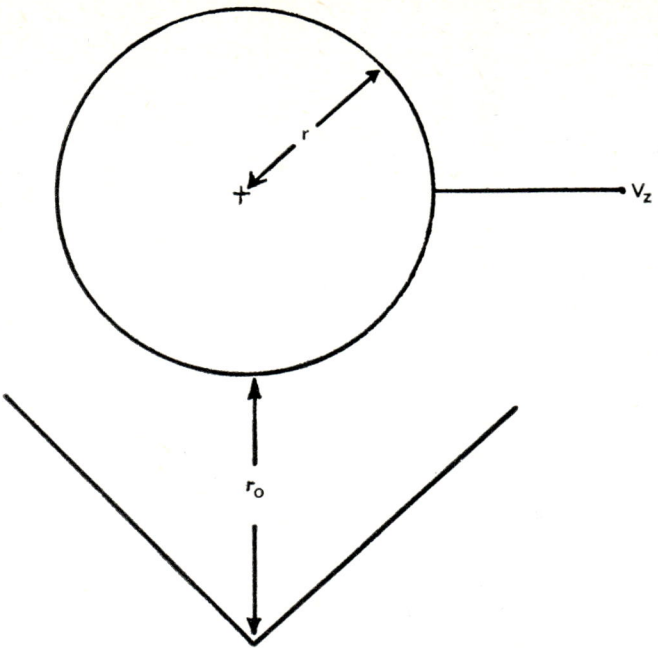

Fig. 1.21. Geometry for a monopole mass filter. r_0 field radius, r rod radius, V_z voltage on rod.

mass selection could be achieved by using only one quadrant of the quadrupole field. He replaced the four rods by a single rod and an earthed electrode (fig. 1.21). The mode of operation is precisely the same and the mass scale is scanned by altering either the frequency or the rod voltages. The behaviour is also very similar, the device being insensitive to ion energy, but providing triangular rather than flat-topped peaks. The monopole has certain inherent advantages: it is easier to construct and align and it does not require such a fine control of the U/V ratio. Further, the ions are resolved in space, and it is possible to detect more than one mass at a time by placing several detectors along the earthed electrode. However, it has the serious disadvantage that it is sensitive to particulate contamination on the rod, and even small irregularities on its surface cause peak splitting and peak height variation. Such a device is therefore not suited to the 'dirty' conditions encountered in day-to-day organic and inorganic analysis, but is eminently suitable for residual gas analysis in ultra-high vacuum systems. A monopole partial pressure analyser has been produced commercially and has the following operating characteristics: mass

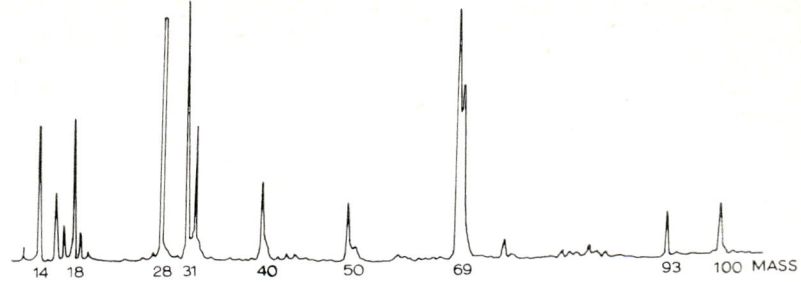

Fig. 1.22. Mass spectrum produced using a monopole mass filter.

scale 1–200, resolution 50, minimum detectable total pressure 5×10^{-9} mm Hg. Fig. 1.22 shows a spectrum obtained with this method.

(c) The quistor

A further variation in the quadrupole geometry was first described in 1959 by Fischer. It was patented as a three-electrode system by Paul and Steinwedel. The three-electrode system is a compact arrangement in which each electrode is a hyperboloid of revolution (fig. 1.23). Using the approximation of circular electrodes, the cross-section of the quadrupole of four rods and the cross-section of a three-electrode assembly of two spheres and a ring are the same. The three-electrode system is, therefore, a three-dimensional equivalent of the linear two-dimensional quadrupole field. Ions travelling in stable orbits do not impinge upon a detector; instead they circulate continuously around

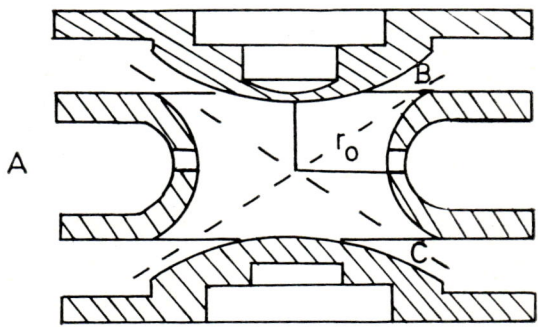

Fig. 1.23. The three electrode system. A quistor. r_0, field radius: A, ring electrode: B and C, upper and lower spherical electrodes.

the axis, and their presence may be detected by the inductive load which they present to an external a.c. source. The three-electrode system thus provides a method of storing ions of a selected m/q value, and has been given the name Quistor. Ions will be lost from the field if they suffer any change of mass or charge, and they may be ejected at will by altering the voltages applied to the electrodes. A number of experimental models have been made and used for studies of ion–molecule reactions and other topics in chemical physics.

1.4. *Methods of ionization*
1.4.1. *Electron bombardment*

In most commercial analytical mass spectrometers, the ionization of gaseous atoms or molecules is carried out using electrons, and this type of source is called the electron bombardment source. The electrons are obtained by thermionic emission from a heated tungsten filament. The tungsten, or more recently rhenium, wire or ribbon is raised to white heat by the passage of an electric current. The electrons emitted into the vacuum around the filament are accelerated to a small electrode (or trap) placed one or two centimetres away, by establishing a potential difference of up to 100 V between the filament and trap.

Fig. 1.24. Three Nier-type ion sources.

The most common potential difference used is 70 V, but dissociation patterns do not change greatly for voltages between 50 and 100 and the collision cross-sections are high, so that sensitivity is good. Below this region the sensitivity gradually falls and the dissociation pattern changes, as the energy in the colliding electron is no longer sufficient for the more energetically expensive processes. Finally, at voltages below 20, it is possible that only ionization and not the subsequent dissociation can occur. It is common practice, therefore, to simplify complex mass spectra by reducing the electron accelerating voltage to a value only slightly greater than the ionization potential of the sample molecules. A further application of this principle is the operation of the combined gas chromatograph–mass spectrometer at an electron accelerating voltage of 20 V. The helium carrier gas, with an ionization potential of 24 V, remains unionized, while any sample molecules with ionization potentials closer to 10 are readily ionized, thus permitting their detection in the presence of larger pressures of carrier gas. The sensitivity of the electron bombardment source is also determined by the numbers of electrons which cross the ionization region between the filament and the trap electrode, or the current which the trap electrode draws from the filament. Currents of up to 500 μA can be achieved in modern analytical machines, but further substantial improvements in electron bombardment sensitivity are not envisaged.

A variation of this type of source is the double electron bombardment source. In this arrangement there is the conventional filament and trap assembly, and the ions are ejected from this region through the ion

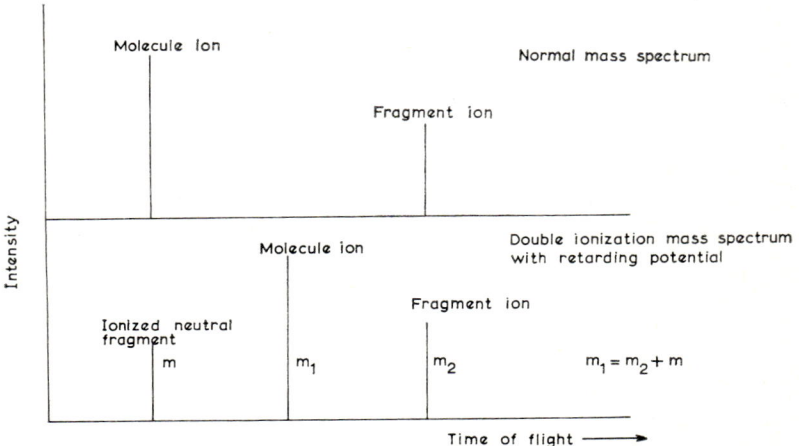

Fig. 1.25. Spectrum using a double ionization source.

beam defining aperture by means of a positively charged backing plate so disposed that the line of flight of the ion is at right angles to that of the electrons. The drawback of the conventional mass spectrometer has always been that it could not detect the unionized fragments of the ion dissociation processes. However, in this modified source the unionized products of the dissociation processes taking place in the first ionization region drift down to a second region with a second filament and trap assembly, where they become ionized and follow the primary ions down to the collector. By operating such a source with and without the second electron beam it has been possible to increase our knowledge of the mechanisms of dissociation processes. An example of a spectrum obtained using this procedure is shown in fig. 1.25.

One of the big disadvantages of the electron bombardment source is the need to employ a white-hot filament. The reader may wonder why the efficient low-temperature electron guns used in thermionic valves and cathode ray tubes are not employed, but such admirable devices are sensitive to contamination and are unsuitable for the ' dirty ' conditions prevailing within the vacuum system of an analytical mass spectrometer. The hot filament is still the most convenient source of electrons although it causes pyrolysis of labile samples (on the filament and on metal surfaces heated by the filament). Some commercial equipment employs water-cooling of the source but this in turn creates problems. There has therefore been a search for alternatives to the electron bombardment source.

1.4.2. *Photo-ionization*

In this method the ionization is carried out by a beam of light (or photons) rather than by a beam of electrons. For such a process to be possible, two factors have to be considered. First, since it is desirable to maintain the sample pressure as low as in the electron bombardment case, the sample must have a high absorptivity. Second, the energy in the photons which strike the sample molecules must be sufficient to carry out the ionization, i.e. this energy must be greater than the ionization potential of the molecule. The energy in light is inversely proportional to its wavelength, and it can be calculated that as the upper limit for the ionization potentials of most molecules is 15 eV, the wavelength of the light required for ionization is 84 nm, which is the far ultraviolet region of the spectrum.

Air absorbs strongly at wavelengths below 185 nm and even the most transparent substance available, lithium fluoride, will not transmit light below a wavelength of 100 nm, so that for photo-ionization the light source must be inside the vacuum system. The simplest light source is a capillary tube along which helium travels from a reservoir to the vacuum system. Electrodes connected to a high voltage supply initiate a discharge in the helium which emits strongly at 59 nm. Even under these conditions, however, the sensitivity is very low because of the low sample pressures, and the photo-ionization source is restricted to special applications.

One such application has grown into a whole new field of spectrometry and is known as photoelectron spectrometry. Here, instead of studying the positively charged product of the ionization process, the secondary electron emitted on ionization is examined. This electron carries with it a good deal of information about the energy levels within the ion. If the light source emits at only one wavelength and this has an energy which is greater than the ionization potential of the sample molecule, then the excess energy is distributed between the translational energy of the electron and the excitational energy of the ion. The ion will have almost no extra translational energy because the velocities of the ion and the electron will be inversely proportional to their masses. There is, therefore, a simple relationship between the energy of the incident photon, the translational energy of the electron, and the excitation energy of the ion. If the translational energy of the electron can be measured, the excitation energy in the ion can be calculated. For any assembly of molecules being ionized, this excitation energy will have

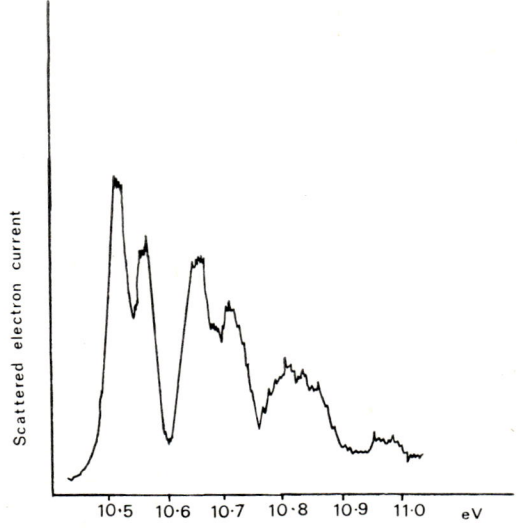

Fig. 1.26. Photoelectron spectrum of ethene.

different values, depending upon the initial and final vibrational and rotational states of the molecule and ion.

By passing the emitted electrons through a hemispherical condenser with defining slits at the entrance and exit and appropriate voltages upon the hemispheres, the translational energy of the electron can be measured. A plot of the numbers of electrons having a specific energy against the electron energy constitutes a photoelectron spectrum and gives details of the electronic and vibrational states within the ion, in the same way as other forms of spectroscopy. It also gives a very precise method of measuring the ionization potential of the molecule free from the ambiguities associated with measurements of ion current. A photoelectron spectrum of ethene is shown in fig. 1.26.

1.4.3 *Chemical ionization*

Ionization of neutral molecules can be induced by collision with other species, such as ions. Ionization of atoms and molecules can also be

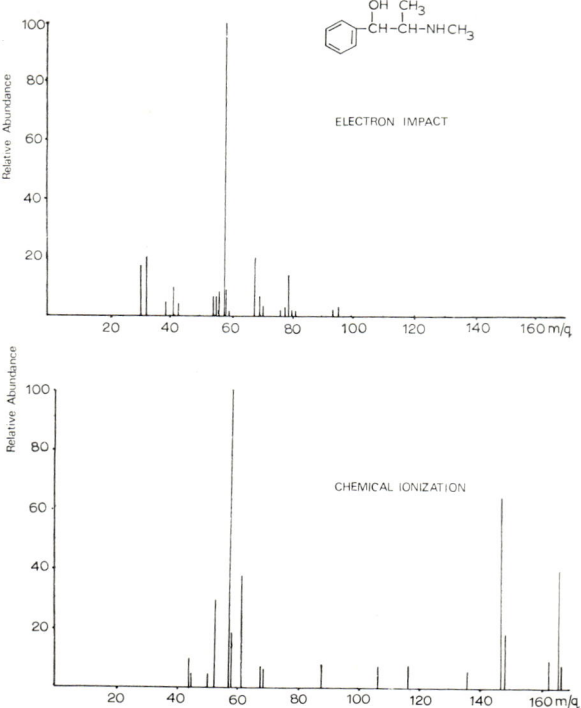

Fig. 1.27. Electron impact and chemical ionization spectra of ephedrine (relative molecular mass 165).

carried out by collision with other charged species, such as positive or negative ions. There is no advantage, however, from the point of view of analytical mass spectrometry, in replacing the beam of electrons by a beam of ions, because of the additional complexity of pumping and electronics. There is, on the other hand, one type of ion–molecule collision which has been exploited in a commercial ionization source and this has been given the name chemical ionization.

If the ion source is redesigned so as to operate at high pressure (up to 266 Pa), then ion–molecule reactions take place rapidly before the ions can leave the ionization region and pass to the analyser. If methane is allowed to flow into an electron bombardment source at high pressure, then the most abundant ion produced is the CH_5^+ ion. If now any substance having a higher proton affinity is mixed with the methane before admission to the source, then proton transfer can take place, as for example:
$$CH_5^+ + C_6H_6 \rightarrow C_6H_7^+ + CH_4.$$

This transfer takes place with high efficiency and leads to a high sensitivity. In addition, the ions and molecules make many collisions, so that the sample ion is not highly excited. As a result, fewer rearrangement processes take place, the chemical ionization spectrum is simpler and the spectra of ephedrine in fig. 1.27 demonstrate the simplification which can be obtained.

1.4.4. *Field ionization*

Finally, there is the most successful of the alternatives to electron bombardment as a means of ionization, the use of extremely high potential gradients.

Consider the change of potential on approaching the outer shells of electrons around the atom. At infinity the potential is zero, but rises slowly until within a few multiples of the atomic diameter, and then begins to rise steeply.

The potential gradient can be calculated from the principles of quantum mechanics and at the outer boundary of most atoms is about 10^8 V cm^{-1}. It is this field which maintains the atom as a stable unit, and if it is modified then the assembly of electrons and nucleus is no longer stable. The potential gradients are modified by the fields of passing electrons, photons and ions, and that is why the atoms become unstable and electrons are given out. However, there is a simpler way to modify the potential gradient at the outer boundary of the atom, by exposing it to the field of an electrode at a very high potential. The electrode requires an edge or point with a very small radius of say 50 nm. A potential of about 10 kV is sufficient to provide a potential gradient close to the edge similar to that at the outermost part of the atom. As a result, any atom drifting near such an edge will have the potential gradient modified and the outermost electron can escape with the formation of an unexcited positive ion. As the sharp edge is positively charged, the ion is repelled and moves to the analytical system of the instrument. The great advantage of this method, which is called field ionization, is that the molecule ion has very little excess energy, so that weak chemical bonds are not ruptured. Thus it is possible to record the mass spectra of very labile biological materials (such as glucose; fig. 1.28) and to examine such compounds as hydrates or alcoholates. Commercial sources tend to use thin wires rather than sharp edges and to increase instrumental sensitivity by encouraging the growth of fine metal whiskers on the wire.

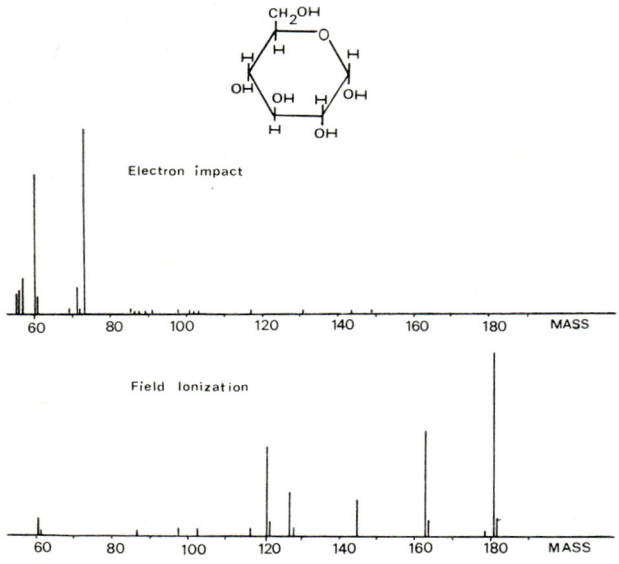

Fig. 1.28. Comparison of the electron impact and field ionization spectra of glucose (relative molecular mass 180).

1.4.5. *Thermal ionization*

A further means of ionization is that of thermal ionization, where the sample is painted on to a filament. Many elements leave the hot filament in the form of positive ions, but this type of source is limited to instruments for measuring isotopic abundance and is therefore not of general interest.

CHAPTER 2
applications to organic chemistry and biochemistry

Most of the great discoveries in organic chemistry before World War II were made without the aid of instruments. The principal tools available were the chemical balance and the thermometer. With the first, chemists carried out the elemental analysis of their samples, and with the second they determined the boiling point or melting point. With such crude characterization they isolated and determined the structures of vitamins, alkaloids, chlorophyll, sugars and terpenes. However, many more recent successes could not have been achieved without the aid of modern instrumentation, and more particularly, modern mass spectrometry. The main reason for this change is that the very great sensitivity of modern mass spectrometers makes it possible to determine the structure of a compound when the amount available is so small that it cannot be seen.

2.1. *Volatility*

In order to study an organic compound using its mass spectrum, it must be possible to produce the compound in the vapour state inside the ion source. For compounds having a boiling point of less than 200° this is quite straightforward; it is merely necessary to allow the vapour to diffuse through a restrictive sinter into the ion source. Compounds of low vapour pressure have to be heated in order to increase their rate of flow through the sinter. Very involatile compounds are heated directly within the ionization chamber, so that they are ionized after evaporation before the molecules can strike a metal surface and condense.

Some samples have such a low vapour pressure that, even under these conditions and at high temperature, no mass spectrum is produced. Alternatively, the temperature at which they would evaporate is higher than that at which they decompose. There are many organic compounds which are too involatile, either because their molecular mass is too high or because they are highly polar, to be studied by mass spectrometry. Compounds of the first type include the proteins, and compounds of the second type include the sugars. Some increase in the volatility of polar compounds may be obtained if they are converted to less polar derivatives. Thus amines may be acylated with perfluoroacyl anhydrides or alcohols converted to trimethylsilyl ethers. However, it is still difficult to produce mass spectra from a compound whose relative molecular mass approaches 2000.

Once in the vapour phase in the ion source, ionization will follow whatever the structure of the organic molecule. The consequences of the ionization of polyatomic molecules may be illustrated by considering the energetics of the process.

2.2. Energetics of ionization

The energy involved in an ionization process is measured by finding the minimum kinetic energy of the electron which is required to carry out that process. The kinetic energy of the electron is determined by the difference in potential between the filament and the trap.

The energetics of the ionization of molecules by electrons may be described most easily in the form of a graph. The simplest possible case is of a diatomic molecule. Its behaviour may be described by a plot of potential energy against interatomic distance (fig. 2.1). The energy rises steeply as the two atoms approach close to one another, due to the repulsions of the bound electrons on the two atoms. As the distance between the atoms increases the energy falls to a minimum and then begins to rise again, as the interaction of the electrons in the bond holding the two atoms together becomes distorted. Finally, as the interaction is weakened, the potential energy rises more slowly and then remains constant with increasing interatomic distance. This corresponds to the situation of two free atoms moving apart; in other words, to dissociation. The diatomic molecule does not in fact behave as a rigid rod with its interatomic distance fixed, but is in a state of oscillation.

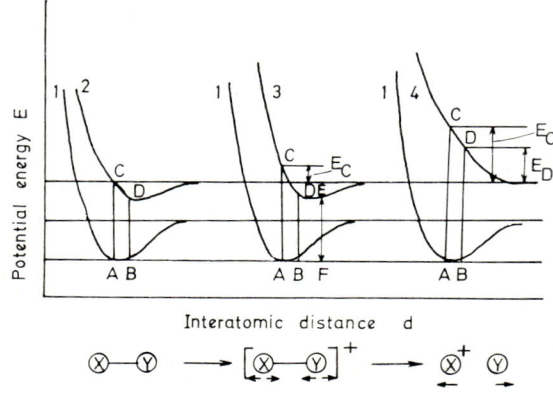

Fig. 2.1. Potential energy curves illustrating the ionization of diatomic molecules.

The atoms vibrate around mean equilibrium positions, the energy associated with these oscillations being quantized just as the energies of excited electronic states are quantized.

In order to represent on the same axes of potential energy and interatomic distance some of the features of the ionization process, the Franck–Condon principle must be invoked. In its simplest form this states that because there is a huge difference in the momenta of the nuclei and the electrons in a molecule, in any process involving the excitation or movement of the electrons there will be no change in either the interatomic distance or the velocity of the nuclei. This means that on the graph all electronic transitions including ionization will be represented by vertical lines. The result of an ionization process will therefore depend on the relative disposition of the curve for the uncharged molecule and the upper curve for the ion. The points A and B on curve 1 represent the nearest approach and farthest separation of the atoms in the ground vibrational level of the diatomic molecule XY. Then, by the Franck–Condon principle, after ionization the interatomic distance must lie between the limits C and D on the upper curves, the points C and D being vertically above A and B. The consequences of the ionization process depend upon the shape and disposition of the upper and lower curves, and the three possibilities are illustrated by the three upper curves. In the first case, as indicated by curve 2, the vertical transition results in the formation of an ion XY^+ which is stable, all the interatomic distances corresponding to discrete vibrational levels of the ion. In the second case, as illustrated by curve 3, the range of interatomic distances between the points C and D now includes the flat portion of the curve, which corresponds to the region where dissociation occurs. Thus in this region the transition results in the formation of an atom X or Y and an ion X^+ or Y^+, but not the ion XY^+. However, the range also includes a region corresponding to some higher vibrational levels of the ion XY^+. The result of ionization is therefore to produce some molecule ions which are vibrationally excited together with some fragment ions and neutral atoms, and these may have relative kinetic energies ranging from zero to the upper limit given by E_C. In the third case, as in curve 4, the range of interatomic distances between points C and D all lie on the flat portion of the curve corresponding to a free ion and an atom. As a result, all transitions give rise to either neutral atoms X and Y or positive ions X^+ and Y^+ moving away from each other with relative kinetic energy between E_C and E_D. There is no chance of producing the stable ion XY^+.

There is thus a relationship between the energy required for ionization, the energy required for dissociation of the bond, and the kinetic energy of the fragments. It could be considered that the ionization energy of a molecule would be that energy required to convert the ground state of the molecule into the ground state of the ion, but this is a theoretical idea which may not be easily measured in practice, because the equilibrium interatomic distance may not be the same for molecule and ion. A more valuable concept is that of vertical ionization potential, which means the energy required to remove an electron from the molecule and take it to infinity without disturbing the interatomic distance. This is represented by the vertical line BD. The difference between the vertical ionization potential BD and the theoretical ionization energy EF is represented for the second case (curve 3) and corresponds to the difference in interatomic distance in the molecule and the ground-state ion. Let A be the energy (called the appearance potential) required to produce from the ground-state molecule XY an ion X^+ and a neutral atom Y moving apart with relative kinetic energy KE. The ionization potential of the atom X is $I(X)$ and it may have some further excitation energy $E(X^+)$. Further, the energy required to break the bond in the molecule XY is $D(X-Y)$. Then the energy required to produce the ion X^+ and the neutral atom Y from the molecule XY is equal to the bond dissociation energy, the kinetic energy of separation of the fragments $KE(X^+,Y)$, the ionization potential of the atom and the excitation energy of the ion X^+ and the atom Y.

Thus
$$A(X^+) = D(X-Y) + I(X) + E(X^+) + E(Y) + KE(X^+, Y)$$

Some at least of these quantities may be measured with the aid of a mass spectrometer. The total energy required for the dissociative ionization process may be estimated by assuming that all the energy of the incident electron is transferred to the molecule, and that none of that energy is transmitted to the ejected secondary electron. The ionization potential of the atom may be determined directly, but is in any case known with great precision from spectroscopy. The excitation energies of the atom Y and the ion X^+ cannot be measured but are usually either zero or so high that they cannot go undetected. The relative kinetic energy can be estimated by finding the kinetic energy of the ion and using the principle of the distribution of momentum between two bodies of unequal mass. In certain cases, therefore, it is possible to get a good estimate of the bond dissociation energy in the molecule. The problem increases in complexity, however, as we move from diatomic to polyatomic molecules. Excitation energies of the fragment ion and atom are less easily identified, and sometimes even the composition of the neutral fragment, which need no longer be a single atom, is in doubt.

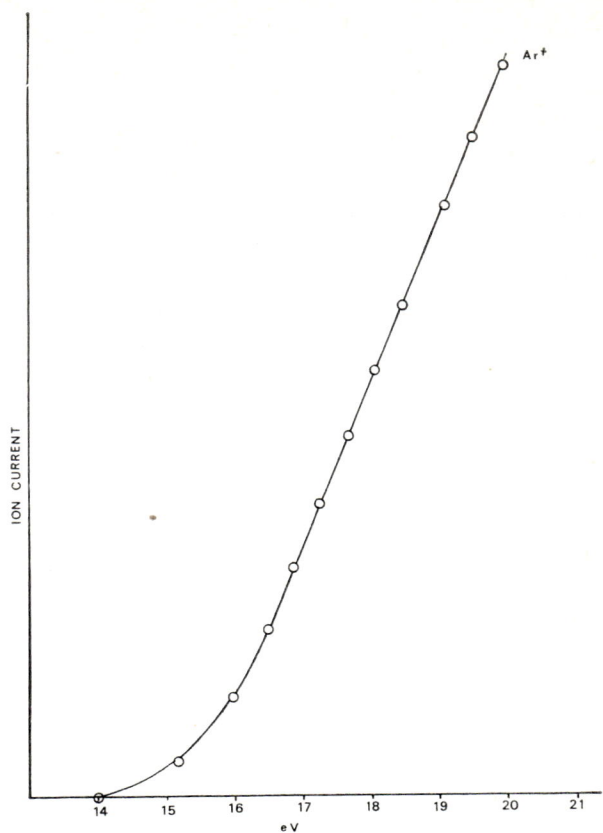

Fig. 2.2.   Ionization efficiency curve for argon.

2.3. *Determination of appearance potentials*

Most mass spectrometers have the means of adjusting the electron accelerating energy by varying the voltage on the trap electrode in the source between 0 and 100 V positive with respect to the filament. The variation in the mass spectrum of a monatomic gas with change in electron accelerating potential is now considered. This is most simply shown by a plot of the singly charged molecule ion current for, say, argon against electron energy in volts as in fig. 2.2. The graph has a rather peculiar shape; the ion current rises above zero at a point on the voltage axis, the rise being exponential in character for a short distance, but then becoming linear.

If there is no excess translational energy associated with either the ejected or the scattered electron, then the appearance potential (A) may be identified as the minimum energy required for the incident electron to allow the formation of positive ions. The calculation of the appearance potential from the ionization efficiency curve is, however, not straightforward. Firstly, because of the exponential nature of the toe of the ionization efficiency curve, it is impossible to detect the exact point at which it meets the axis. Secondly, whatever the measured electron voltage at which this occurs, it certainly does not correspond to the true value of the appearance potential, which in this case is equal to the ionization potential of argon.

These two difficulties have different causes. The first is due to the fact that in a conventional ion source the electrons are emitted from a white-hot filament, and so have a distribution of thermal energies superimposed upon the kinetic energy imparted by the electron accelerating potential. Thus, as we approach the toe, the ionization is increasingly being caused by those electrons having the higher thermal energy; and these are of course fewer in number as the amount of thermal energy increases. For this reason the curve approaches the axis asymptotically. The second problem is caused by the fact that electrons do not necessarily experience the potential which is applied to the plate, and which may be read by a voltmeter, because the nature of the surface of the trap electrode may change by adsorption of gas molecules, so creating what is called a contact potential. The contact potential will, however, to a first approximation be the same for all molecules. All that is therefore required is to calibrate the voltage axis, by recording at the same time the ionization efficiency curve for a gas whose ionization potential is known spectroscopically, for example krypton.

The first problem has interested mass spectrometrists for many years. One solution is to treat the data obtained from the curves either graphically or mathematically, so as to eliminate the effect of the thermal distribution. An alternative solution is to construct an ion source which has a monoenergetic beam of electrons. The mathematical solutions range from the simple device of plotting the logarithm of the ion current against electron energy, to the complex deconvolution process by which a computer is used to disentangle the variation in ion current caused by the change in voltage from that caused by the variation in thermal velocity. The instrumental solution ranges from the comparatively simple device of using a retarding potential upon a grid to select those electrons above a certain defined velocity, to the more complex electron monochromators, which consist of pairs of charged hemispheres, between which the electrons pass and are segregated according to their velocity.

Whatever method is used the end product is a more or less precise value for the appearance potential, which is the minimum energy required for the ionization process. If the target gas is atomic then A corresponds to the ionization potential of the atom, and good agreement with spectroscopic values is generally obtained. If the target gas is a molecule and it is the ionization efficiency curve of the molecule ion which is being studied, then A corresponds to the ionization potential of the molecule. Once again, agreement with other values is generally good, although there are sometimes discrepancies when there are excited states lying closely above the ground state of the molecule ion. An example of this is benzene, and the values obtained have ranged from the generally accepted value of 9·28 V to above 10 V.

When the ionization efficiency curve being studied is that of a fragment ion, then before it is possible to make even an approximate estimate of the bond dissociation energy in the molecule, the ionization potential of the atom or radical must be known. If, for example, the appearance potential of the methyl ion from methane is measured, the ionization potential of the methyl radical must be known before the upper limit to the C–H bond dissociation energy can be calculated.

$$CH_4 + e \rightarrow CH_3^+ + H^{\cdot} + 2e$$

It is an upper limit because the appearance potential A may involve a component E which is the sum of the relative kinetic energy of the methyl ion and the hydrogen atom and the excitation energy of the methyl ion. The value of $D(CH_3-H)$ may be calculated using published values for A and I. The calculated bond dissociation energy D is close to that derived from gas kinetic studies, so that in this case the excess energy E is small.

$$A(CH_3^+) = D(CH_3-H) + I(CH_3) + E$$

This method of measuring bond dissociation energies is sometimes called the electron impact method.

There are certain other relationships which can be used if the thermochemical data are available. Thus the energy required for the dissociative ionization, that is the appearance potential A, is equal to the difference between the enthalpies of formation of the products and reactants.

$$A(X^+) = \Delta H_f(X^+) + \Delta H_f(Y) - \Delta H_f(XY)$$

If the enthalpies of formation of XY, X and Y are known the ionization potential of X can be calculated. Alternatively, knowing $I(X)$ the enthalpy of formation of the radical or atom can be calculated.

Measurements of the strength of chemical bonds and the ionization potentials and enthalpies of formation of molecules, using studies of the variation of ion current with electron energy, become less reliable as the relative molecular mass increases. From a structural standpoint it may be assumed that the ionization process results in the formation of a highly excited molecule ion, and that the mass spectrum is a record of the chemical reactions which the ion undergoes. Some attempts have been made to predict the mass spectra of simple organic molecules and the most successful of these attempts is called the 'quasi-equilibrium theory'. In this theory a population of ions is assumed to be formed by Franck–Condon processes and to have a distribution of excitation energies. The excited ions have an appreciable lifetime and this is sufficient to allow the distribution of the excitation energy throughout the modes of vibration of the ion. The accumulation of a sufficient amount of this excitation energy in one mode of vibration can lead to the rupture of a chemical bond. There is thus formed an equilibrium distribution of excited ions whose rate of dissociation or bond rupture is related to the amount of excitation energy which they have acquired from the incident electrons. The rate constants (k) for dissociation are derived by considering each ion as a set of n simple harmonic oscillators. k is then proportional to $[(E-E_0)/E]^{n-1}$, where E is the total excitation energy and E_0 is the activation energy required for dissociation. The probability of an ion acquiring an excitation energy E is calculated separately. While this theoretical treatment is of interest to the physicist, a more empirical approach is suited to the needs of the organic chemist. It is sufficient to consider the mass spectrum as resulting from a series of competing and consecutive unimolecular decompositions of excited ions and to identify some of the dissociation pathways.

2.4. *Interpretation of the mass spectra of polyatomic molecules*

When a molecule containing only a comparatively few atoms is introduced into the ion source of a mass spectrometer, the mass spectrum obtained under normal operating conditions is surprisingly complex. The first stage in the extraction of chemical information from a mass spectrum is the classification of the types of ion responsible for the peaks in the spectrum.

2.4.1. *Molecule ion peaks*

If the sample consisted only of a monatomic gas, for example argon, the bulk of the ion current generated would be due to the molecule ion of argon at mass 40, and the mass spectrum would consist substantially of one large peak. Nevertheless this simple spectrum would contain one of the most important pieces of chemical information that the method can provide: the relative molecular mass of the sample substance. At this stage only the integral mass is being considered although much more information can be obtained from a study of the precise mass. It is rare that the peak of highest mass in a mass spectrum of a polyatomic molecule is in fact the molecule ion peak, because of the isotopic complexity of most naturally occurring elements. It is, however, usually easy to distinguish the molecule ion peak due to the most abundant isotopic species, as outlined in the next section.

For most organic compounds the molecule ion or parent peak will have an even mass and an odd number of electrons. An exception is provided by compounds containing one nitrogen atom, such as ammonia (mass 17) or pyridine (mass 79). This is also true for compounds containing any odd number of nitrogen atoms, but not for compounds containing an even number, as for example hydrazine (mass 32) or dinitrobenzene (mass 168). Elements having an even mass and an odd valency are rare, so the appearance of an odd numbered molecule ion peak in the mass spectrum of an organic compound suggests that the compound contains nitrogen. The proportion of the total ion current carried by the molecule ion, as revealed by the height of the parent peak, varies widely with the structure of the molecule ion. When the molecule contains an atom having a high electron affinity, the direct formation of the molecule ion requires more energy than that required for the formation of a positive fragment ion and a negative atomic ion. As an example, the molecule ion of tetrafluoromethane is never formed, although it is quite possible to measure the ionization potential of the molecule by photo-ionization. The electron impact process results in the formation of the trifluoromethyl positive ion and the fluoride negative ion:

$$CF_4 + e \rightarrow CF_3^+ + F^- + e.$$

It is a general rule that molecules containing halogen atoms tend to have mass spectra with comparatively small molecule ion peaks, and in many cases such peaks are absent. The relative importance of the molecule ion peak diminishes with increasing relative molecular mass as in ascending an homologous series. Unsaturation appears to stabilize the molecule ion, as shown by the example of the compounds hexane and hex-1-ene. In the former, the molecule ion gives a peak in the mass spectrum which is 14·1 per cent of the base peak (or most intense peak), while in the latter it is 28·6 per cent. Another feature promoting stability in the molecule ion is the occurrence within the molecule of aromatic structures. The molecule ion peak of cyclohexane amounts to only 70·5 per cent of the base peak, while in the case of benzene it is the preponderant peak. In the case of large polycyclic hydrocarbons almost the whole of the ion current is carried by the molecular ion, as

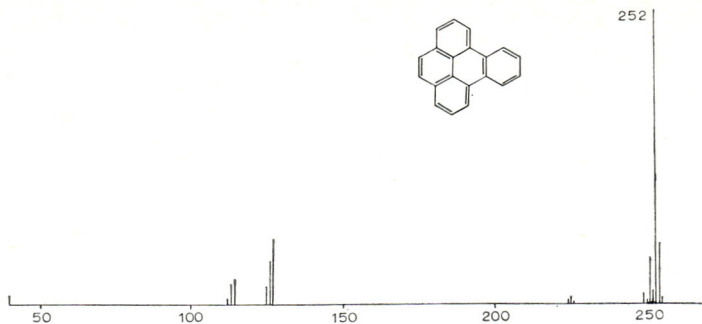

Fig. 2.3. Mass spectrum of the carcinogenic polycyclic hydrocarbon benzo(*e*)pyrene (relative molecular mass 252).

can be seen in fig. 2.3. The case of hexafluorobenzene is an interesting one since the effects described oppose one another. Despite the high fluorine content, the molecule ion carries a substantial amount of the total ion current.

Substances which are thermally unstable may pyrolyse on heated surfaces within the ion source before ionization, and so give very small molecule ion peaks in their spectra. Such compounds will however show the expected spectrum if they are ionized in a water-cooled source or in a field ionization source.

2.4.2. *Isotopic peaks*

In the mass spectrum of an organic compound, the peak with the highest mass value is not in agreement with the molecular mass measured by some other physical technique, such as ebulliometry. The peak may have an odd mass value even when the compound does not contain nitrogen. Further, at one mass unit lower there is a much larger peak. The reason for the appearance of this pair of peaks is the polyisotopic nature of naturally occurring carbon. All naturally occurring carbon contains 1·1 per cent of ^{13}C, and this is why on the mass scale of ^{12}C = 12·000000 the relative atomic mass of ordinary carbon is 12·011. The polyisotopic nature of natural carbon has nothing to do with radioactive carbon-14, which is formed by the passage of cosmic rays through the atmosphere and provides the basis for carbon-14 dating. The amounts of this isotope are so small that they are not normally detected by mass spectrometric means. Thus all compounds which contain carbon will exhibit this characteristic pair of peaks corresponding to the ^{12}C and ^{13}C molecule ions. The ratio of the heights of these two peaks will vary with the number of carbon atoms in the molecule, since each carbon

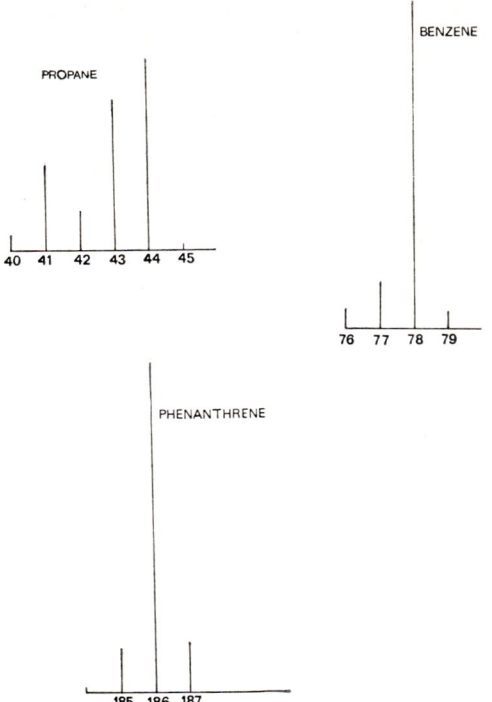

Fig. 2.4. Effect on the mass spectrum of the polyisotopic character of naturally occurring carbon.

atom can provide a ^{13}C molecule ion with an abundance of 1·1 per cent. The spectra of a series of carbon compounds are given in fig. 2.4. From the ratio of the two peak heights at m and $m+1$ it is possible to deduce the number of carbon atoms in the molecule.

A number of other commonly occurring elements are polyisotopic in nature; oxygen contains ^{17}O and ^{18}O, nitrogen contains ^{15}N, and hydrogen contains deuterium. The natural abundances of these are too small to be of use diagnostically, but artificial enrichment with the heavy isotopes can lead to the use of the mass spectrometer in tracing techniques.

Sulphur-34 is, however, sufficiently abundant naturally to allow the identification of sulphur in a compound by the observation of a pair of peaks at m and $m+2$ with a peak height ratio of 20 : 1. The most characteristic isotope patterns are those exhibited by the halogens chlorine and bromine. Any molecule containing a chlorine atom will

51

mass analysis. When the compound is both unknown and impure, a frequent problem with material extracted from biological sources, it becomes difficult to disentangle the spectrum of the major component from those of impurities. The most unfavourable situation is when several unidentified compounds are present in about the same concentration. It is then necessary to make some prior separation of the individual compounds before recording the mass spectra. Sometimes, when large samples are available, for example in the examination of petroleum fractions, conventional methods for the isolation of pure organic compounds can be used. Fractionating columns with high separation efficiency up to 100 theoretical plates may be used in the purification of volatile liquids, while crystallization or partial freezing can be employed to provide pure samples of solid compounds. However, the mass spectrometrist is rarely provided with such an abundance of material, because under such circumstances conventional methods of analysis might be more appropriate. Usually sample sizes are in the milligram to microgram range and special methods for separating mixtures on this scale are required.

Chromatographic methods, in which the components of a mixture are distributed repetitively between two phases, have the advantage that the efficiency of separation is frequently enhanced by reduction in the sample size, and so these methods are most suitable for the pre-treatment of sample mixtures. The simplest example to understand is perhaps thin-layer chromatography. Here the components of a mixture are distributed between a liquid solvent phase and a solid adsorbent phase. This distribution is best illustrated by the familiar isotherm in which the concentration of a single substance in the solid adsorbent is plotted against the concentration in the liquid phase. Such a graph could be plotted from the results of a number of equilibrations of solutions of different concentrations with the same amount of solid adsorbent at the same temperature (see fig. 2.8). It can be seen that at high concentrations of the sample the adsorbent becomes saturated. In order to get satisfactory separations of mixtures, the concentration of each component must be in the range covered by the linear part of the isotherm. Two conditions are necessary for satisfactory separations: (*a*) the slopes of the isotherms must be sufficiently different, and (*b*) on reducing the concentration the same isotherm must be obtained as that with rising concentration, that is to say there is no irreversible adsorption. The principle of the separation may be understood by taking an extreme example in which for one component no adsorption takes place. The concentration in solution remains constant and the isotherm is therefore a vertical line; the second component is insoluble and so its

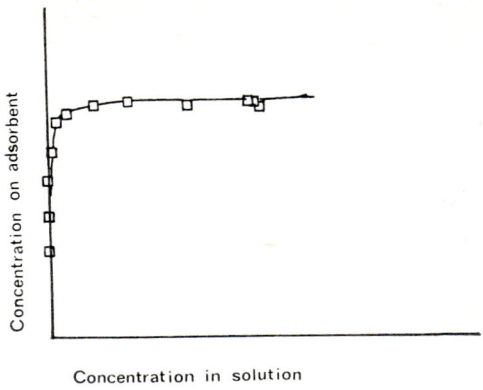

Fig. 2.8. Adsorption isotherm.

isotherm is a horizontal line. The separation of such a pair of components could be achieved in a single equilibration, but for all real mixtures separation depends on less extreme differences in the slopes of the isotherms, and so a single equilibration gives only a partial separation, in which one phase has a greater concentration of one component and a smaller concentration of the other than has the second phase. Effectively complete separation can be achieved by multiple equilibration, in which the solution is successively presented to fresh samples of adsorbent while the adsorbent loaded with sample is successively presented with fresh amounts of solvent.

Such a procedure would be cumbersome and time-consuming if it could not be accomplished automatically. However, the simple device of keeping the adsorbent in the form of a column or thin layer and flowing the solvent through it satisfies the requirement.

2.5.1. *Thin-layer and paper chromatography*

In the technique of thin-layer chromatography the solid adsorbent, which is most commonly silica gel, is coated as a thin layer on a glass plate. The sample mixture is applied in solution as a small spot close to the edge of the plate. After drying, the plate is placed in a glass tank which contains a solvent or solvent mixture to a depth just below the spot position. The solvent rises through the thin layer by capillary action and carries with it the components of the mixture. The rate at which each component rises, and the ultimate height of each component when the solvent front reaches the top of the plate, is determined by the distribution coefficients, or slopes of the respective isotherms. If the isotherms are linear, the original spot will separate into a series of spots of approximately the same size. In practice, the spot will increase in size slightly because of lateral diffusion, and irreversible adsorption will be revealed by tailing of the spots.

The presence of the spots may be demonstrated by a variety of techniques. If the components of the mixture are not themselves coloured, they may be converted to a coloured derivative with a suitable reagent. Alternatively, the adsorbent may be loaded with a fluorescent compound and the plate examined under ultraviolet light. Quenching of the fluorescence by the components results in their appearance as dark spots against a uniformly glowing background. Once the positions of the spots have been fixed, the area of adsorbent covered by each spot is carefully scraped from the plate and extracted with a suitable solvent. The solvent is then evaporated and the residue transferred to the mass spectrometer for identification.

In such an experiment, illustrated in fig. 2.9, a mixture of aromatic hydrocarbons was separated on a thin layer of cellulose acetate and the mass spectra of the separate components recorded. The total quantities of each hydrocarbon were less than one microgram. Apart from the obvious need for care and precision in the manipulative procedures, this method suffers from the disadvantage that few adsorbents are so pure

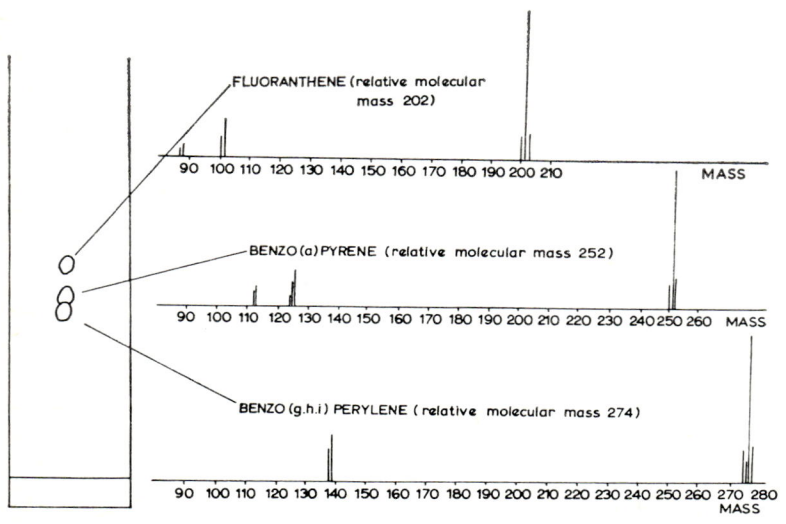

Fig. 2.9. Combination of thin-layer chromatography and mass spectrometry for the identification of polycyclic hydrocarbons.

that extraction by organic solvents will not remove organic material from them. The mass spectrum arising from this material will be recorded together with that of the component. This drawback is shared by the companion techniques of column and paper chromatography. In the latter case, the process of partition is a complex one, involving not only the cellulose fibres of the paper itself but also the film of liquid on the surface, which may not have the same composition as the flowing solvent phase. Once again, although well defined spots are obtained and good separations achieved, identification is made difficult by the extraction of a variety of organic compounds from the paper during the elution of the components.

2.5.2. *Gas–liquid chromatography*

The most sophisticated chromatographic method yet devised was invented by Martin. In this technique, partition takes place between the gas or vapour phase and the liquid or solution phase, and therefore it is restricted to the examination of volatile substances. It is also referred to as gas chromatography or vapour phase chromatography.

The principle is formally the same as for other chromatographic methods. A volatile substance has a finite vapour pressure, so that in a closed vessel at any one temperature there exists a fixed concentration of the substance as vapour in the space above the liquid sample. If the pure substance is replaced by a solution in an involatile solvent, there will still be a finite, though different, concentration of the substance existing as vapour in the space above the solution. This will depend on the concentration of the solution, the attractive forces between the molecules of the sample substance (as indicated by the enthalpy of vaporization) and the attractive forces between the molecules of the solute substance and the molecules of the solvent (as indicated by the enthalpy of solution). For ideal solutions the vapour pressure of the solute, or the concentration in the vapour phase, will be proportional to the concentration of the solute in the involatile liquid—as expressed in Raoult's Law. If the single solute is replaced by a pair of solute substances A and B, and either their boiling points or their enthalpies of solution or both are different, then the ratios of the concentrations of A and B in the vapour and liquid phase will be different. That is to say, their partition coefficients between the solution and vapour phase will be different. Once again it is this difference in partition coefficient which allows the separation of volatile substances.

Repetitive batch equilibration is not feasible and must be replaced by a continuous process, but the problem is that of transport between successive equilibrations. In the case of thin-layer and paper chromatography the transport is provided by the moving solvent under the influence of capillary forces. In the case of column chromatography the transport is provided by the solvent moving under the influence of gravity. The transport of the vapour of the sample components in gas chromatography can only be achieved by the movement of an inert gas stream.

In order to maximize the area of the interface across which the equilibrations must take place and cut down the time for the establishment of equilibrium, the involatile solvent must be presented to the flowing inert gas in the form of a thin film. If a pure solvent is exposed to solute vapour, the rate at which solution takes place depends upon the probability of the solute molecules striking the solvent liquid surface and this probability will be proportional to the surface area. Once the solute molecules are in solution, equilibration will be complete only when they have diffused to all parts of the solvent and the concentration is uniform. The rate at which this state of affairs is achieved depends on the volume of the solvent, or, for a fixed surface area, on the thickness of the solvent layer. All these considerations were embodied in Martin's original design, in which a stream of nitrogen gas carried the volatile components of a mixture through a narrow tube filled with a packing of granular solid material with a high surface area. This was coated with a thin film of the involatile liquid, dinonyl phthalate.

A small sample of the mixture was injected into the carrier gas stream as a plug of vapour and passed along the tube, suffering repeated partitions between the gas phase and the liquid phase on the granular support. The components of the mixture travelled along the tube at different rates, depending on their relative volatility or affinity for the involatile liquid, and emerged one at a time from the end of the column. A detector capable of distinguishing between the carrier gas and the component indicated the emergence of each component from the column. The original detector was a sensitive gas density balance. Since then many detector systems have been designed, but the one most commonly used today is based on flame ionization. In this system the carrier gas leaving the base of the column is injected into a small hydrogen/oxygen flame which surrounds a pair of electrodes. The electrodes are connected to a power supply and a means of detecting any small current carried by the flame. When only the carrier gas passes through the flame, no current is detected, but when a component of the mixture which contains carbon passes through the flame, ions and free electrons are formed and the flame conducts, so providing a signal to the detector. The detector output is fed to a recorder and the final trace is a record of the rise and fall of the partial pressure of each

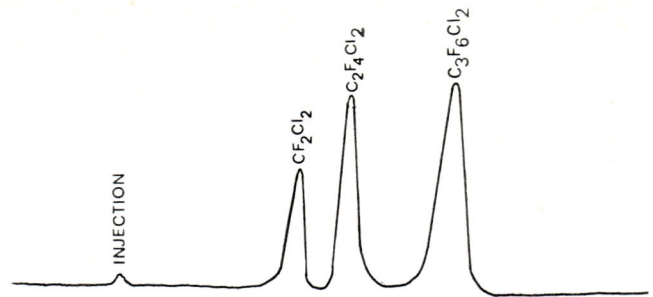

Fig. 2.10. A gas chromatogram of halogenated alkanes.

component in the mixture as it passes through the detector. The record is called a chromatogram, and fig. 2.10 shows a chromatogram of some halogenated alkanes.

The gas chromatograph separates the components of a mixture in time and in space and provides the ideal complement to the mass spectrometer, more so because the sensitivity of the two instruments is about the same. It is possible to isolate each component of a mixture by trapping it in a tube cooled in liquid air as it leaves the base of the column, although a by-pass must be used so that only a fraction is burnt in the flame ionization detector. The tubes, together with the trapped components of the mixture, can then be transferred individually to the inlet system of a mass spectrometer. This procedure is very time-consuming and leads to losses of the components, so that it would clearly be advantageous to combine the separation on a gas chromatograph and identification on a mass spectrometer.

A simple solution would appear to be the direct coupling of the base of the column to the inlet system of the mass spectrometer, but this is not feasible. The main reason is sensitivity, for at best the effluent carrier gas contains only one per cent of the component.

Some prior concentration of the sample component is therefore required, and this can be achieved most simply by substituting helium for nitrogen as the carrier gas and making use of its high rate of diffusion. The effluent helium carrier gas from the base of the column is passed through a porous glass tube, the outside of which has a vacuum jacket connected to a pumping system. The helium passes through the walls of the porous tube and is pumped away, while the components diffuse along the porous tube to the ion source of the mass spectrometer. The result is that the mass spectrum is substantially that of the component and not of the carrier gas. Other devices which have been used include membranes permeable to organic compounds but not to helium, and jet separators, which use the difference in momentum of helium atoms and the larger organic molecules. A further improvement may be achieved by ionizing the effluent with a voltage less than the ionization potential of helium and greater than that of most organic molecules, for example 20 V. Whatever device is used, the gas chromatograph–mass spectrometer combination provides a method of obtaining the mass spectra of pure samples of organic compounds from a mixture which may contain up to 100 components. Further, the total sample size needed is less than 1 μg, for this method is very economical of sample. One final requirement is that the spectrum should be recorded in a fraction of the time taken for the component to emerge completely from the column, so that very high rates of scan of, say, 5 seconds must be attainable.

If the substance separated is a known one, and its mass spectrum has been recorded previously, it may be identified by spectrum matching. Further, the concentration of the substance may be deduced by comparing the area of the gas chromatographic peak with that obtained using a known concentration of an authentic pure sample.

If the compound is not known or spectrum matching has been unsuccessful then the structure must be established from first principles. Some information can be obtained from the low resolution mass spectrum, including the relative molecular mass, and the presence of atoms having distinctive isotope patterns. A complete determination of the composition may be made from a study of the high resolution spectrum. The atomic constitution of each ion responsible for a peak in the spectrum may be determined from its precise mass. This is an excellent check on the efficiency of the purification process, and gives some idea of the functional groups existing within the molecule. Confirmation of these ideas is often achieved by carrying out a chemical modification of the unknown substance and studying the change in the mass spectrum. An example of this method is given by the identification of the heterocyclic compound shown in fig. 2.11. The precise mass of the molecule ion shows that its atomic constitution is $C_6H_6O_3$. Addition of heavy water to the sample causes a change in the mass spectrum. Hydrogen has been exchanged for deuterium and the molecule ion peak has moved up one mass unit. So have the other peaks corresponding to ions derived from that part of the molecule. It can be concluded that there is only one OH group in the molecule. A second oxygen atom may be involved in a carbonyl group and so the sample is treated with hydroxylamine to form the oxime. The new spectrum shows a peak 15 mass units higher, confirming the presence of one carbonyl group. In this way the structure of this molecule, which is a plant alkaloid precursor, was determined.

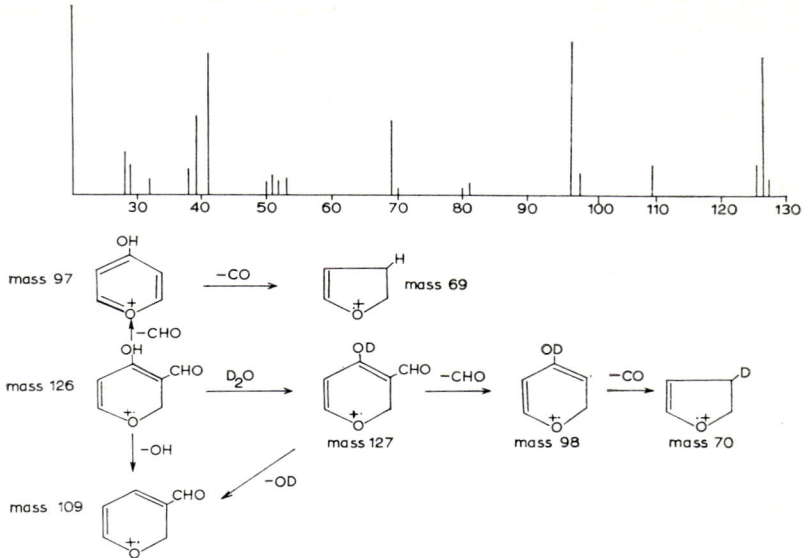

Fig. 2.11. Identification of an alkaloid precursor by a combination of mass spectrometry and deuterium exchange.

2.6. *Some examples*
2.6.1. *Glycerides*

Although the relative molecular masses of many oils and fats are high, and some exceed 1000, most of them are sufficiently volatile for study by mass spectrometry. They have been known for many years to be esters of propane-1,2,3-triol (glycerol) with long chain fatty acids and are called triglycerides. The mass spectra are particularly informative, and all triglycerides give a spectrum containing a molecule ion peak which can be measured with sufficient precision to identify its atomic constitution. As an example, an unknown oil from Northern Iraq was identified by its mass spectrum with a molecule ion peak at 878 as a highly unsaturated triglyceride. The most intense peak in the spectrum is that due to an ion formed by the loss of an acyloxy (RCOO·) fragment from the molecule ion. Where the triglyceride is formed from glycerol and three different fatty acids, there are three such fragment peaks each corresponding to the loss of one of the acyloxy groups. As an example, the spectrum (fig. 2.12) of 1-myristo-2-stearo-3-palmitin (1-oxycarbonyltridecane-2-oxycarbonylheptadecane-3-oxycarbonylpentadecane propane) has a molecular ion peak at mass 806 and three fragment peaks at mass 579 (loss of $C_{13}H_{27}COO·$), mass 551 (loss of

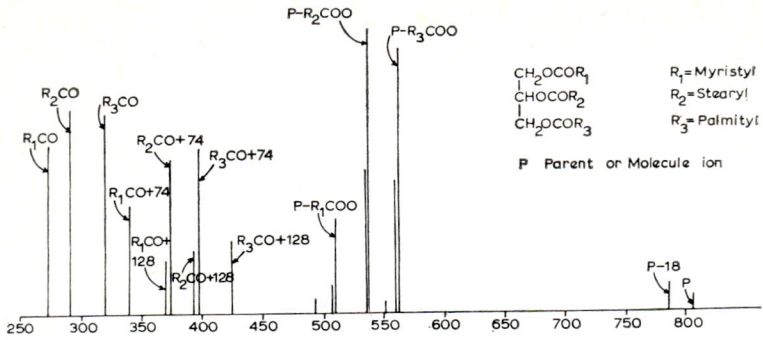

Fig. 2.12. Mass spectrum of the triglyceride 1-myristo-2-stearo-3-palmitin.

$C_{15}H_{31}COO\cdot$) and mass 523 (loss of $C_{17}H_{35}COO\cdot$). The three acids are also represented in the spectrum as peaks due to the acyloxy (RCOO·) and acyl (RCO·) groups. Thus the determination of the structure of a triglyceride of a series of straight chain acids is a comparatively simple matter. More complex triglycerides may require prior chemical treatment such as hydrolysis to assist in the elucidation of their structure.

2.6.2. Plastoquinones

There is an interesting class of natural compounds called plastoquinones which are found in the chloroplasts of plants, where they perform functions associated with electron transport. They are most commonly isolated from alfalfa. The mass spectra of all members exhibit the same fragment ion peak at mass 189 corresponding to the atomic constitution $C_{12}H_{13}O_2$, which has been identified as a derivative of 2,3-dimethyl-1,4-benzoquinone (2,3-dimethylcyclohexa-2,5-diene-1,4-dione). This portion is the active site in the molecules, and it is the reduction of the quinone to the semiquinone or hydroquinone ions in the chloroplast which provides the electron transport. The residue of the molecular structure differs from compound to compound, but is in all cases derived from a long chain of 2-methylbuta-1,3-diene-(isoprene) units.

These units are common building blocks in plant molecular architecture and are incorporated into the terpenes (the constituents of the fragrant woods) and into natural rubber. The polyisoprenoid skeleton in plastoquinones carries different substituent groups for different members of the class. In the simplest member, with a molecular mass of 748, the atomic constitution given by the precise mass of the molecule ion is $C_{53}H_{80}O_2$, and the mass spectrum shows a series of peaks due to ions produced as a result of the loss of an increasing number of isoprene C_5H_8 units from the side-chain in the molecule ion. Confirmatory evidence has been obtained by reducing the compound to the corresponding benzene-1,4-diol (hydroquinone) and acetylating it. The mass spectrum of this new compound is in accord with the structure of the original given below.

<chemical structure: 2,3-dimethyl-1,4-benzoquinone with side chain $-(CH_2-CH=C(CH_3)-CH_2)_9H$>

In addition to this simple compound, there are other plastoquinones in which the isoprenoid side-chain is substituted with hydroxyl or ester groups.

2.6.3. *Peptides*

One of the most exciting discoveries in the mass spectrometry of organic compounds was the observation that polypeptides were sufficiently volatile to permit their introduction into the ion source. These compounds are similar in structure to the proteins, although with a much smaller relative molecular mass. They have the general formula:

$$NH_2\text{-CH(R}_1)\text{-CONH-CH(R}_2)\text{-CO}\ldots NHCH(R_n)COOH$$

In order to increase the volatility it is common practice to esterify the terminal carboxyl group and acetylate the amino group to give derivatives of the type RCONHCO ... COOR′. From the mass spectra obtained, the molecular weights and atomic compositions can be established for polypeptides of up to 15 amino acid residues with a sample size of only a few micrograms.

From a study of the spectrum of a naturally occurring peptidolipid (that is a polypeptide in which the terminal amino group is acylated with a long-chain fatty acid residue), it was realized that the spectrum contained sufficient information to reveal the amino acid sequence within the peptide. Before this, sequences had always been determined by the classical methods of partial hydrolysis or stepwise degradation, and even though these steps could be automated and the amino acids identified by chromatographic methods, the new method had the advantage of speed and extreme sensitivity. The compound which provided this breakthrough was a peptidolipid aptly named fortuitine, for by a happy chance it occurred as a pair of homologues, the long chain N-acyl groups differing by two CH_2 groups. This had the effect that every peak in the mass spectrum corresponding to an ion containing the acyl group appeared as a pair of peaks 28 mass units apart, with the peak heights in the ratio of the relative amounts of the two homologues in the original sample. In the preparation obtained by the extraction of *Mycobacterium fortuitum*, the two forms were derived from eicosanoic and docosanoic acids. The mass spectrum showed a pair of peaks due to the molecule ions at 1331 and 1359, corresponding to the atomic constitutions $C_{70}H_{125}N_9O_{15}$ and $C_{72}H_{129}N_9O_{15}$. These atomic constitutions can be interpreted as arising from the methyl ester of a nonapeptide with the terminal amino group acylated by either a C_{20} or a C_{22} carboxylic acid. There are characteristic fragment ions corresponding to the loss of two acetyl groups suggesting that two of the amino acids are also acetylated. However the most significant part of the spectrum was the intense series of peaks resulting from the rupture of the peptide bonds (NHCO). These peaks revealed a stepwise loss of one amino acid residue after another from the molecule ion, each fragment ion being clearly identified by a pair of peaks 28 mass units apart, since the N-acyl groups were still intact. The structure of fortuitine was determined by this means as

$C_{19}H_{39}$–CO–Val–MeLeu–Val–Val–MeLeu–Thr–Thr–Ala–Pro–OCH_3
                                                     | |
($C_{21}H_{43}$–)   Ac   Ac

where Val = valine, MeLeu = methylleucine, Thr = threonine, Ala = alanine, Pro = proline and Ac = acetyl.

Since the first demonstration of amino acid sequencing by mass spectrometry, a considerable number of polypeptides have been studied and their structures elucidated. The most recent trend has been to attempt sequencing using a low resolution mass spectrometer and a computer to abstract the relevant information, for in addition to the main mode of decomposition which has been discussed there are a number of other modes of decomposition for such complex molecule ions. It is no longer necessary to have the polypeptides with a long acyl group, and the computer programme is able to determine the sequence by matching characteristic losses from the molecule ion with the masses of known amino acids or their derivatives.

2.6.4. *Macrocyclic pigments*

One of the more surprising observations is the comparative ease with which macrocyclic pigments such as the phthalocyanines, the chlorophylls and the porphyrins may be volatilized. They are all large but compact molecules, consisting of four pyrrole rings linked together by nitrogen atoms or CH. groups into 16-membered rings. The metal atom at the centre may be copper as in the industrial dye Monastral Blue, magnesium as in chlorophyll, or iron as in the haem of haemoglobin. Different members of each group may carry a variety of side-chains. The molecule ion peak is always the most intense (as shown in fig. 2.13), and an unusual feature of the spectra is the appearance of a doubly charged molecule ion peak. The dissociation patterns obtained are easily rationalized in terms of the loss of the pendant groups from the molecule ion. While most of the porphyrins have had their structures determined by purely classical chemical degradation, there is a recently studied group, the so-called petroporphyrins obtained from petroleum, which have been investigated with the mass spectrometer. This group of pigments has been shown to be derived from the same basic macrocyclic structure but with different alkyl side groups.

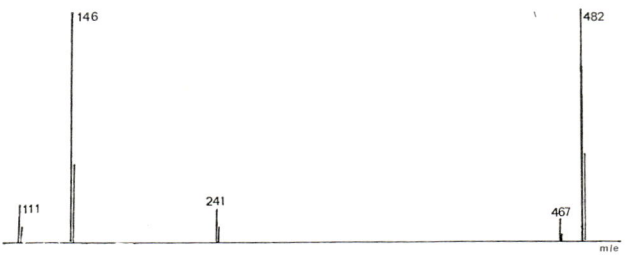

Fig. 2.13. Mass spectrum of a macrocyclic pigment—nickel octamethyltetraaza porphyrin ($C_{24}H_{24}N_8Ni$).

CHAPTER 3
the mass spectrometer in inorganic chemistry

3.1. *General considerations*

The compositions of inorganic substances can be studied using the spark source mass spectograph, or the electron bombardment mass spectrometer. The nature of the sample and the information required dictate which technique is most suitable. Involatile samples such as refractory oxides, e.g. zirconia, can be examined directly only by incorporating them into graphite electrodes and sparking them in the source system of the mass spectograph. On the other hand a volatile liquid halide such as arsenic trichloride can be evaporated directly into the vacuum system of an electron bombardment mass spectrometer. This section reviews the information about the nature of inorganic substances which can be obtained by the second of these two techniques, the first being dealt with later.

If an electron bombardment ion source is used, then either the inorganic substance must have a sufficient vapour pressure at the temperatures normally accessible, say up to 500°C, or the source must be modified to include a high temperature furnace. In the Knudsen cell furnace, the sample species evaporating are ionized before striking surfaces within the source. Many elements are sufficiently volatile to permit examination in one or other of these ways, but in general, metals are introduced in the form of their hydrides or halides, or in the form of a complex with such species as carbon monoxide, nitrogen oxide, cyclopentadiene or any of a large number of chelating agents.

The low resolution mass spectrum can provide a considerable amount of information about the structure of such volatile inorganic compounds. The relative molecular mass can be determined and this may reveal the state of polymerization of the compound. Molecule ion or fragment ion peaks may be multiplets with a characteristic pattern of peak heights corresponding to the naturally occurring isotopic constitution of an element. A high resolution study can then provide details of the atomic constitution of all ions, and this, coupled with the dissociation pattern, can frequently give structural details. A measurement of the variation of relative peak height with temperature may allow a determination of the heat of sublimation or the heat of dissociation of polymers. A record of the variation in ion current with electron energy gives thermochemical data, such as ionization energies and heats of formation and, in favourable cases, bond dissociation energies. This type of information may be difficult to obtain in other ways. Finally, the presence of isomers may be detected by recording the variation in peak height with time during the course of an evaporation.

3.2. *The Knudsen cell*

In the Knudsen effusion cell involatile species are heated in a furnace within the vacuum system. The furnace has a small exit aperture whose area is negligible with respect to the total internal surface area. The sample leaving the cell by the aperture is in equilibrium with that within the furnace, and travels as a molecular beam to the ionization region of the mass spectrometer. The ion current representative of any particular species is then proportional to the vapour pressure of the species within the furnace. The Knudsen cell furnace is constructed of some refractory material such as molybdenum, tantalum or alumina, and may be heated most simply by electron bombardment from tungsten filaments situated nearby to temperatures as high as 2500°C. The temperature is usually read by an optical pyrometer through small holes drilled in the metal radiation shield which must surround the furnace. The sample is most commonly placed in the cell before evacuation and heating, but may be created by a chemical reaction within the cell by introducing a second gaseous species, as in the preparation of ruthenium oxide from ruthenium and oxygen gas.

3.3. *Metal chelates*

In the early nineteen-twenties Sir Gilbert Morgan, who has been described as the father of modern inorganic chemistry, performed some pioneering investigations in his laboratories in Birmingham University. He studied the complexes of metals with 1,3-diketones such as pentane-2,4-dione (acetylacetone). He was astonished, not only by the remarkable stability of these complexes, but also by their volatility. He announced his discovery, that metal acetylacetonates could be distilled without decomposition, in the poetic words: "Acetylacetone has given wings to the metals". A large collection of his original samples still exists in the museum of the Chemistry Department, and fig. 3.1 shows a selection of these. As a link with the past, the mass spectrum of one of his original samples has been recorded in the same Chemistry Department where the sample was prepared half a century ago. Fig. 3.1 also shows the spectrum and reveals the high state of purity of the compound. In the generally accepted structure of this type of compound the metal atom is almost enclosed within an organic shell, and such compounds are called chelates (from the Greek *chelos*, a crab's claw).

The metal derivatives of diketones can be passed through a gas chromatography column, thus opening up the possibility of separating

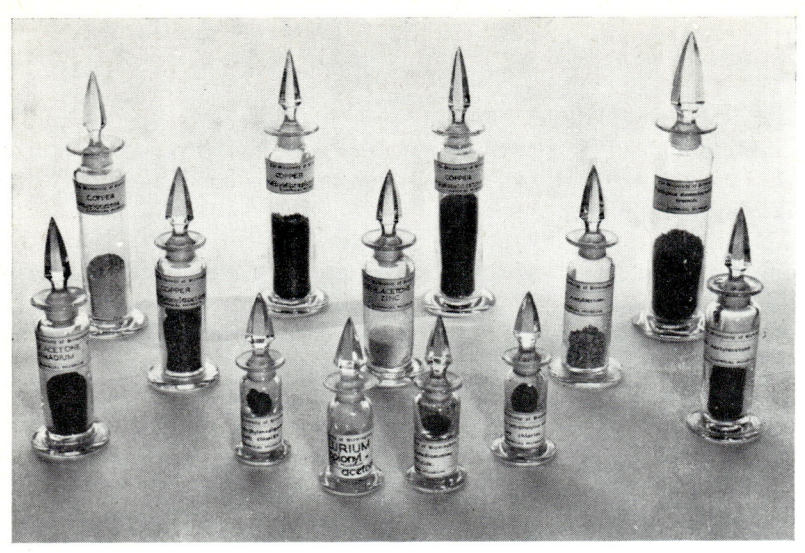

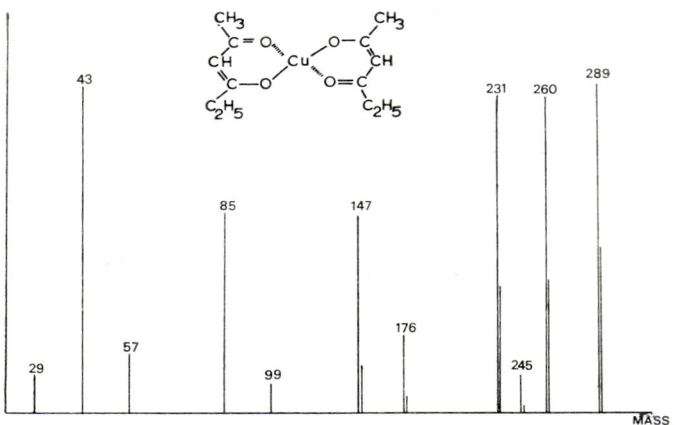

Fig. 3.1. (a) A collection of the earliest metal chelates made by Sir Gilbert Morgan and (b) a mass spectrum of a sample of one of them, copper propionylacetonate.

metals of closely similar chemical properties. There is a very great increase in volatility if the methyl groups of acetylacetone are replaced by bulky branched alkyl groups or by perfluoroalkyl groups, and the oxygen atoms replaced by sulphur or nitrogen. Very many of these compounds have been prepared and their mass spectra studied. Some of the compounds, e.g. chromium trifluoracetylacetonate, are so volatile that they may be admitted to the ion source at room temperature.

This class of compounds has some interesting general features which have been demonstrated by mass spectrometry. All metal derivatives of diketones and ketoesters are capable of undergoing reactions during their evaporation, and these reactions may be classed as reactions of exchange, polymerization, and association. The exchange reaction may be demonstrated by simultaneous evaporation of two species, $CF_3COCHNaCOCH_3$ and $C_3F_7COCHKCOC_4H_9$. The resulting mass spectrum will reveal not only the peaks originating from these two pure compounds, but also peaks coming from two new compounds in which the alkali metal atoms have been exchanged. It is not so easy to demonstrate whether the second reaction, polymerization, takes place on evaporation or merely reflects the structure within the solid phase. However, the mass spectra of many metal acetylacetonates contain peaks due to multiples of the molecular structure, for example the spectrum of the compound CaL_2 (where $L = C_3F_7COCHCOC_4H_9$) contains peaks due to the species $Ca_3(C_3F_7COCHCOC_4H_9)_6{}^+$ and $Ca_2(C_3F_7COCHCOC_4H_9)_4{}^+$. The relative peak heights appear to vary with temperature, suggesting the direct evaporation of polymers within the crystal lattice. On the other hand, coevaporation of species containing different metal atoms results in the appearance of polymeric ions containing both metals, and it is possible that both mechanisms are involved.

Association reactions are simpler to identify. If one evaporates a mixture of an alkali metal and alkaline earth metal derivative of a diketone, the resulting mass spectrum contains peaks due to species which involve two metal atoms and three diketone residues. Similar association reactions have been shown to occur with alkali metal derivatives of diketones and the corresponding lanthanide derivatives. Such compounds as $KHo(C_2F_5COCHCOC_4H_9)_4$ have been shown to exist and, in some cases the mass spectrometric results have been confirmed by X-ray crystallography. These compounds can exist in two isomeric forms. If the lanthanide derivative of a diketone is examined by following the change in ion current with time for a chosen peak, say the molecule ion peak, a smooth curve of Gaussian form is obtained if the diketone is symmetrical. When the diketone is unsymmetrical, two separate peaks are obtained for the two possible isomers. This isomerism is explained by considering that the diketone molecules bridge the corners of a coordination octahedron, as in fig. 3.2. The case is somewhat simpler if the metal atom is in a planar environment when

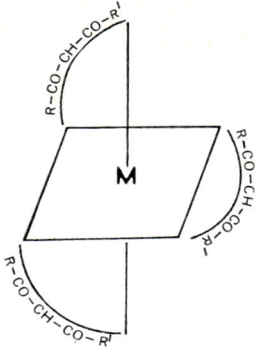

Fig. 3.2. Isomerism of metal chelates derived from an unsymmetrical 1,3-diketone and a trivalent metal.

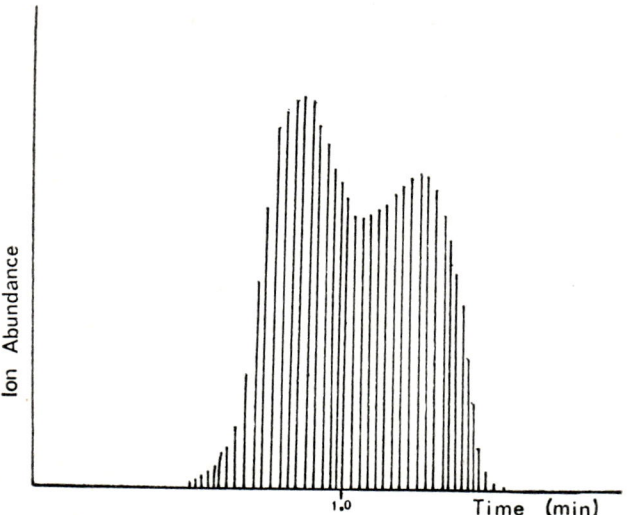

Fig. 3.3. Record of the variation of the molecule ion current with time during the evaporation of a small sample of zinc oxinate.

only *cis–trans* isomerism can occur as in the two forms of $(CH_3CSCHCOCH_3)_2Pb$.

It is not only the metal acetylacetonates which are sufficiently volatile to be studied in this way. In fact many other metal chelates can be examined. A common organic reagent which gives precipitates with most metal ions is 8-hydroxyquinoline or oxine. While higher admission temperatures have to be used, the resulting spectra are remarkably simple, consisting in general of the molecule ion peak and the fragment ion peaks caused by the successive loss of the oxine residues. Once again, because oxine is an unsymmetrical molecule, it has been possible to demonstrate the existence of the isomers of the trivalent metal derivatives, such as aluminium oxinate, and those of the planar divalent metals such as zinc. So far this has not been achieved by the application of any other technique (fig. 3.3).

Perhaps the most familiar example of a metal chelate encountered in school laboratories is the bright red precipitate produced by mixing a solution of a nickel salt with a solution of butane-2,3-dione dioxime (dimethyl glyoxime). The characteristic colour has been used as a test for nickel ions for over 50 years. Only recently, however, was it shown to be sufficiently volatile to permit the recording of its mass spectrum. It behaves quite differently from other chelates, in that it does not take part in exchange reactions in the ion source. When a butane-2,3-dione dioxime is mixed with the nickel derivative of a different 1,2-dioxime and evaporated into the ion source, the resulting mass spectrum is merely the superposition of the spectra of the two pure compounds. This is intriguing because if the nickel ion is precipitated in the presence of a mixture of the two chelating agents and the resulting mixed precipitate evaporated, the mass spectrum now reveals the presence of an additional chelate in which the nickel atom is bonded to one of each of the dioximes (fig. 3.4). Similarly, palladium also gives volatile chelates with dioximes, but when the nickel and palladium derivatives of two different oximes are coevaporated, no exchange of metal atoms takes place. The complexes described all have a square planar configuration and if an unsymmetrical dioxime, such as phenylpropane-2,3-dione dioxime, is used, the two possible isomers may be separated by solvent extraction. However, the presence of these isomers cannot be shown by any structure in the curve of molecular ion current with time, because at a temperature of 150°C the two forms become interconvertible and only a broad peak results. Nickel forms a chelate with the closely related compound 2-hydroxybenzenecarbaldehyde oxime (salicylaldoxime) and the mass spectrum of this compound can also be recorded. In this case, however, the possibility of the existence of two isomeric forms is clearly demonstrated by recording the variation of the molecular ion current curve with time, two distinct peaks being revealed (fig. 3.5).

At present there is renewed interest in the study of metal chelates because of the possibilities they afford for the extraction of metals without causing atmospheric pollution. Metals may be dissolved from

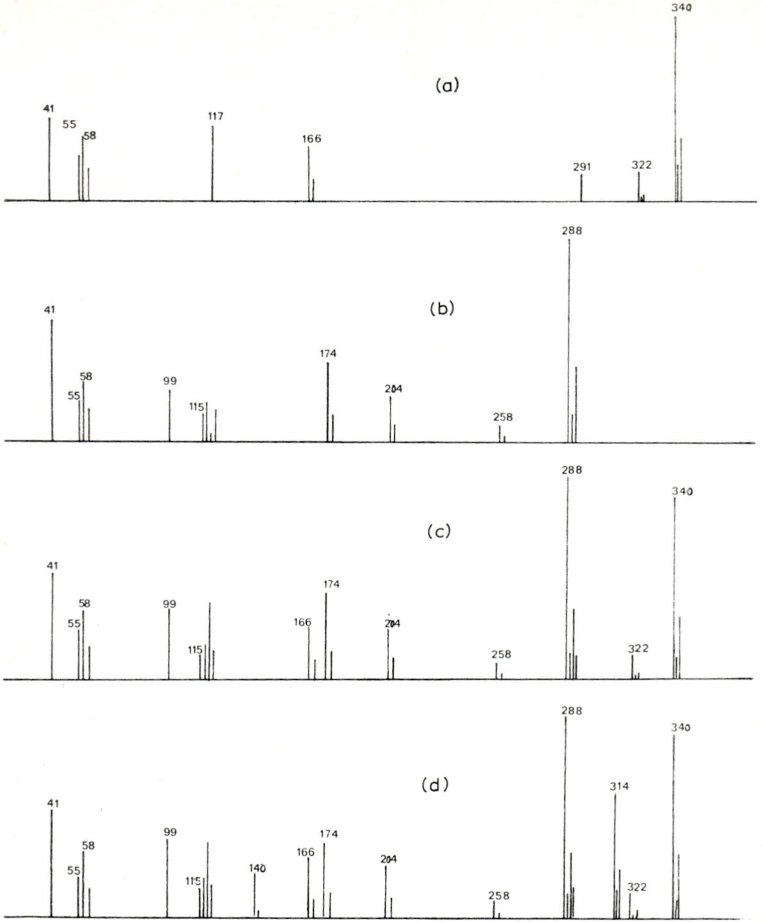

Fig. 3.4. Investigation of the precipitation of the nickel ion in the presence of two different dioximes using mass spectrometry. (a) Nickel cyclohexane--1,2-dione dioximate. (b) Nickel butane-2,3-dione dioximate. (c) Mixture of (a) and (b) co-evaporated. (d) Chelate obtained by precipitation of the nickel ion with a mixture of butane-2,3-dione dioxime and cyclohexane-1,2-dione dioxime.

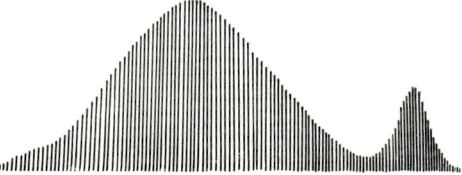

Fig. 3.5. Demonstration of the existence of *cis–trans* isomers of metal chelates. The molecule ion current rises and falls twice because of the differing rate of evaporation of the two isomers.

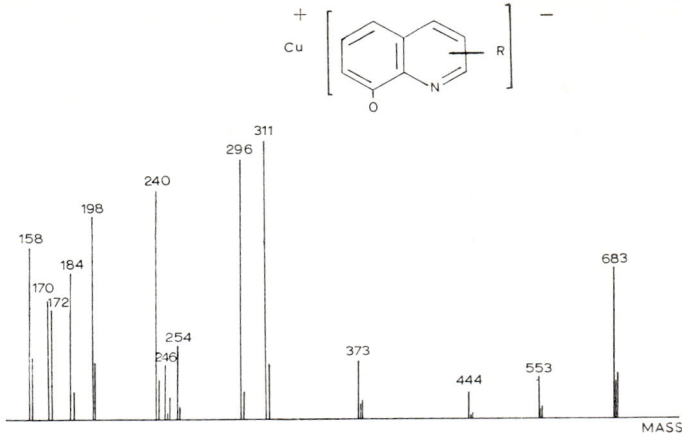

Fig. 3.6. Mass spectrum of a metal chelate derived from copper and a commercial metal extractant. The precise structural formula of this commercial compound is not known.

ore or scrap with acid, extracted into petrol or some other cheap solvent with a specific metal chelate, and the pure metal may be stripped from the organic phase, which is then returned for recycling. Fig. 3.6 shows the spectrum of the copper derivative of a commercially available chelating agent. This is based on 8-hydroxyquinoline, but there is a long chain hydrocarbon substituent to confer solubility in petroleum.

The application of mass spectrometry to general inorganic chemistry can be considered under two headings. First, the Knudsen cell can be used to examine refractory materials, permitting both a determination o their structure and the equilibria which exist between species at high temperatures. Second, the mass spectrometer can be used to examine volatile metal compounds such as hydrides, or organometallic compounds such as carbonyls and alkyls.

3.4. *High-temperature studies*

The halides of alkali metals are examples of polar compounds, and it is strange to think of their existence in the gas phase. However, because of the high vacuum within a mass spectrometer and its ability to detect very small vapour pressures, it is possible to record the spectra of alkali metal halides at quite modest source temperatures. From the present concepts of the three-dimensional lattice structure of alkali halide crystals it is less surprising that, when this structure breaks down on heating, it does not provide only diatomic units, but a range of polymers.

Thus the mass spectra of lithium halides have peaks corresponding to the ions of monomeric, dimeric and tetrameric species, the dimeric peaks being most significant in the case of lithium fluoride. Probably because of decreasing volatility, the intensities fall with increasing mass of the halogen and alkali metal, and only dimers are observed for caesium chloride and iodide. The heat of dissociation of dimers to monomers has been calculated to be about 120 kJ mol^{-1}. When mixtures of alkali metal halides are evaporated there is some formation of mixed polymeric ions containing both metals, e.g. LiNaI$_2$. Although lithium oxide mainly sublimes by dissociation, there is evidence of the existence of Li$_2$O among the gaseous species. A mixed oxide of lithium and sodium (NaOLi) was discovered from a high temperature mass spectrum of lithium oxide containing sodium as a contaminant. It would be difficult to identify such a compound by any other method. Even the hydroxides of the lower alkali metals are sufficiently volatile to permit the recording of their mass spectra. The spectrum of lithium hydroxide indicates that, while the most abundant species is the monomeric LiOH, dimers and trimers do exist. Sodium hydroxide, on the other hand, evaporates mainly as the dimer. Studies on alkali metal alkyls show that they are highly associated, ethyllithium occurring in the gas phase as a hexamer and ethylpotassium as a tetramer. Further structural evidence from the dissociation pattern of ethyllithium hexamer suggests that the molecule consists of an inner group of lithium atoms surrounded by ethyl groups.

The mass spectra of all the metals of group II have been recorded, and the heat of vaporization of zinc and cadmium have been determined from the change in the peak height with temperature. The spectra of the fluorides of the lighter elements show evidence of the existence of dimers and trimers, but this is not observed for the heavier members. A particularly successful study has been made of the ionization and appearance potentials of the dichlorides of magnesium, calcium, strontium and barium. The heats of formation and the bond dissociation energies so calculated are in excellent agreement with thermochemical data.

Several studies have been made of group II metal alkyls. Dimethylberyllium, which is highly polymeric in the solid state, exists as the monomer in the gas phase, although there are low intensity peaks arising from the octamer. The dialkyl derivatives of zinc, cadmium and mercury are all monomeric, and the metal–carbon bond dissociation energies have been determined by electron impact.

The high temperature facilities of the mass spectrometer have been used in a number of studies on the evaporation of the group IV elements, carbon, silicon, germanium and lead. The results were unexpected, showing in addition to the monatomic species, polyatomic molecules with up to seven metal atoms in the molecule. Because the ionization potentials of the elements were well known, it was possible to calculate bond dissociation energies in such species as Sn_4 and Ge_6. The molecules are believed to be linear and in one case, that of C_3, this is known to be true from details of the C_3 emission spectra obtained in smoky flames and in the heads of comets. The carbon–carbon bond dissociation energy is greater in its triatomic molecule and in the corresponding pentatomic molecule C_5, but falls for those molecules containing even numbers of carbon atoms, a fact which suggests some change in structure, at present unexplained. This behaviour was not shared by other group IV elements. From measurements, during vaporization, of the ion currents for the different polyatomic species, it was possible to calculate not only the heats of reaction for the depolymerization process, but also the heat of sublimation of the different polyatomic species, and this was extended to mixed species containing more than one kind of element. The vapour of silicon carbide was shown, for example, to contain the species SiC, Si_2C, Si_2C_2, Si_2C_3 and Si_4. Evaporation of mixtures of germanium, silicon and carbon revealed the existence of molecules such as $GeSiC$ and Ge_2SiC.

From early times it has been known from simple vapour density measurements that sulphur is associated in the vapour above molten sulphur, and many textbooks claim that the major species is S_8 and not S_2. In fact a mass spectrometric study reveals the existence in vapour of all the species S_n where n varies from one to eight, and even small quantities of S_9 and S_{10}. On the other hand, evaporation of sulphur from a specific allotropic form produces homogeneous vapour. Thus the free evaporation of rhombic sulphur, which is believed to be largely composed of S_8 rings, gave rise to a vapour consisting substantially of S_8 molecules. Sublimation of Engel's sulphur, which is believed to be composed of S_6 units, gave only S_6 molecules. Recently, new allotropic forms have been prepared and shown to be S_{10} and S_{12}. The equilibrium distribution of all the polyatomic sulphur molecules is re-established if any of these allotropic modifications is mixed with alumina before heating, and it is assumed that this substance acts as a catalyst for the interconversion of the allotropes. Similar polyatomic species have been identified in the vapour of selenium, and the strengths of the bonds holding the atoms together have been determined, together with the heats of interconversion of some of the polyatomic molecules. Heating mixtures of sulphur and selenium together at temperatures above 200°C leads to the formation of octatomic molecules, in which all possible proportions of sulphur and selenium atoms are represented. Other mixed molecules which have been identified include S$_7$Te and SeTe.

3.5. *Volatile inorganic compounds*
3.5.1. *Boron hydrides*

There are comparatively few derivatives of group III elements which are sufficiently volatile to permit their examination by mass spectrometry, but among them are the strange and fascinating boron hydrides. These compounds, which appear to defy the simple rules of valency, with such formulae as B_2H_6, B_5H_9 and B_4H_{10}, nevertheless have a formal resemblance to the aliphatic hydrocarbons. They are low boiling and form a similar series of molecules containing chains of atoms. They are, however, sensitive to moisture and air, and the mass spectrometer has proved to be one of the most successful ways of studying them. Although the resulting spectra have been compared with those of the alkanes, the spectra are in fact quite different, those of all boron compounds having a pair of peaks at the mass values of 10 and 11, corresponding to the two naturally occurring boron isotopes. However, there are peaks at almost every mass number from 10 upwards to the mass value corresponding to the molecule ion containing ^{11}B. The problem of identifying the separate boron hydrides is made more difficult by the fact that ^{10}BH and ^{11}B have closely similar masses and high resolution is required to separate the individual peak components. Once the peak identities are established, it is possible to get structural information about the boron hydrides.

The simplest but most elusive member of the series, BH_3 or borane, was first positively identified by mass spectrometry. Its existence had been predicted as an intermediate in the thermal reactions of the boron hydrides, and it was found among the pyrolysis products of diborane (B_2H_6). This molecule contains two hydrogen atoms more than would be expected from simple valency considerations, and so it is not surprising that its spectrum has some unusual features. It is obvious, in fact, that two hydrogen atoms are in a different environment compared with the other four. At the present time it is believed that these two hydrogen atoms form a bridge between the two boron atoms, as in the structure below:

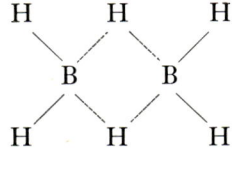

The strength of this bridge, or to put it more precisely the bond dissociation energy $D(BH_3-BH_3)$, can be measured in two ways in a mass spectrometer: by using a Knudsen cell and measuring the change in degree of dissociation with temperature, or by measuring the ionization potential of the borane molecule and the appearance potential of borane from diborane, assuming no thermal dissociation of the diborane during the measurements. The results give a value of 2·56 eV or 258 kJ mol^{-1}. This means that the bond is about as strong as a carbon–bromine bond. The structural unit was shown to be:

$$\begin{array}{c} H \\ \diagdown \\ B\!-\!-\!B \\ \diagup \\ H \end{array} \begin{array}{c} H \\ \diagup \\ \\ \diagdown \\ H \end{array}$$

This structure was also observed in tetraborane B_4H_{10}. In the identification of this compound a sophisticated use was made of the isotopic constitution. Not only were the tetraboranes prepared from isotopically pure ^{10}B and ^{11}B, but specifically labelled molecules were created in which either the terminal or an inner boron atom was an isotope different from the other three. In addition, the hydrogen atoms were replaced either totally or partially by deuterium atoms.

Investigations of the pyrolysis of tetraborane revealed the existence of a number of other higher boron hydrides. An unstable intermediate, whose existence had been suspected, was shown to be B_4H_8, and among the more stable products were identified B_5H_9, B_5H_{11}, B_6H_{12}, B_7H_{11}, B_7H_{13}, B_8H_{12}, $B_{10}H_{14}$ and B_9H_{15}. The most stable of these compounds is the pentaborane B_5H_9, and this is reflected in the mass spectrum by the high intensity of the molecule ion peak. The remainder of the spectrum is in accord with a cyclic structure. The second pentaborane B_5H_{11} is much less stable, and the molecule ion peak is vanishingly small. At the temperature of boiling water it decomposes into borane (BH_3) and tetraborane B_4H_8. Among the higher boron hydrides the most interesting is decaborane $B_{10}H_{16}$. Its structure is symmetrical, being composed of two B_5H_8 units derived from pentaborane by the loss of a hydrogen atom from the apex of the pentagonal prism. The two units are thus connected by a boron–boron bond, and the strength of this bond has been measured by electron impact to be 3·2 eV or 307 kJ mol^{-1}. Thus the B–B bond is appreciably stronger than that of diborane, and this is in accord with both the dissociation patterns and with quantum mechanical equations. Boron hydrides with relative molecular masses up to 230 have been detected ($B_{20}H_{26}$), but the elucidation of their structure is still at an inconclusive stage.

3.5.2. *Group IV hydrides*

The hydrides of group IV elements are also highly volatile and (apart from the hydrocarbons) require careful handling to avoid decomposition. The mass spectra of all the analogues of methane have been recorded, including plumbane PbH_4, and more complex hydrides, such as disilane Si_2H_6 and digermane Ge_2H_6, have been identified, as well as some mixed hydrides which are even less familiar.

Because of their similarity to organic hydrocarbons, considerable attention has been paid to the chemical reactions of group IV hydrides, and the mass spectrometer has been extremely useful in the analysis of the products of photochemical and radiolysis studies of silane and germane. As an example, the attack on silane with methyl radicals has been examined kinetically, and the products shown to be methane and the silyl radical SiH_3. Other compounds of this type which have been investigated include the tetra-alkyl and tetra-aryl derivatives of silicon, germanium, tin and lead. Perhaps the most interesting of these is tetraethyllead, which is used as a petrol additive to impart anti-knock properties and aid in fuel economy. The toxic nature of many lead compounds has led to a campaign against the use of tetraethyllead in petrol, and this in turn has required the development of methods capable of analysing very low concentrations of this and other similar organometallic compounds in air. At present the only technique capable of such an analysis is based on the mass spectrometer.

3.5.3. *Group V compounds*

The hydrides of group V elements are claimed to be extremely poisonous. Recently there was a tragedy at sea in which a number of seamen were overcome by the fumes of arsine, AsH_3, leaking from a cylinder in the hold. For this reason, and because of the sensitivity of phosphine and its analogues to moisture and air, much of the information about their chemistry has been gained using the mass spectrometer. This requires only minute quantities, and it may be coupled directly to the vacuum line in which the hydrides are prepared. The mass spectra of all the simple hydrides have been reported and, in addition to the simple hydrides, compounds containing two and three atoms of the element have been identified, such as P_2H_4 and P_3H_5. These are both fairly unstable and, on heating, P_2H_4 has been shown to break down into P_2H_2, phosphine (PH_3), and phosphorus.

Mass spectrometry has been used to study the hydrolysis of mixtures of the calcium and magnesium phosphides, arsenides and antimonides. When they are hydrolysed separately, compounds such as phosphine, arsine, stibine and their homologues are formed. In the gas mixture prepared by a simultaneous hydrolysis the following compounds were observed: PH_2NH_2, NH_2PHNH_2, PH_2PHNH_2, PH_2PHPH_2, PH_2AsH_2, AsH_2NH_2, NH_2AsHNH_2.

Another interesting compound whose chemistry has been studied is N_2F_4. The N–N bond dissociation energy in this molecule is so small, about 80 kJ mol^{-1}, that at quite moderate temperatures the most abundant chemical species is NF_2. This is a free radical (cf. NO_2) and adds to unsaturated hydrocarbons. The study of the kinetics of such reactions is facilitated by the mass spectrometer's ability to measure the NF_2 radical concentration directly.

3.6. *Volatile organometallic compounds*

The electronic structure of the transition metals permits the formation of a wide variety of organometallic compounds, and these have provided many problems for the structural inorganic chemist. Fortunately, a common characteristic of these substances is their volatility; and this has allowed the mass spectrometer to be used in the elucidation of some structures, the simplest being the metal carbonyls.

The transition metal carbonyls have been known for many years, and their low boiling points prompted early workers to use them in studies of the isotopic constitution of the elements. J. J. Thomson attempted to use nickel carbonyl in his positive ray apparatus but was unsuccessful, the compound decomposing on the walls of the vessel. Aston, however, was able to use the carbonyls of iron, chromium and nickel in his mass spectrograph.

Interest was reawakened after the last war with the growth of structural studies in inorganic chemistry, and the spectra of most metal carbonyls have now been recorded. In the ionization of $Ni(CO)_4$, $Fe(CO)_5$, $Cr(CO)_6$, $Mo(CO)_6$ and $W(CO)_6$, the electron which is removed comes from an orbital closely associated with the metal atom. In the subsequent dissociation neutral carbon monoxide molecules are removed one by one, the appearance potential rising by a large and increasing increment as each CO is removed and the ion becomes smaller. The average energy of dissociation for a single CO loss, as determined by electron impact, is in good agreement with that obtained calorimetrically. More complex metal carbonyls can be prepared in which two metal atoms or more are involved in the molecule. The spectra of dimanganese decacarbonyl $Mn_2(CO)_{10}$ and the analogous compounds of rhenium and technetium $Re_2(CO)_{10}$ and $Tc_2(CO)_{10}$ have been recorded. The metal–metal bond in these compounds is broken both thermally and by electron impact. The radical $Mn(CO)_5$ may be generated pyrolytically from the decacarbonyl. From measurements of its ionization potential and the appearance potential of the ion $Mn(CO)_5^+$ from $Mn_2(CO)_{10}$, the metal–metal bond dissociation energy has been estimated, and has been shown to be very weak, only about 1 eV or 100 kJ mol^{-1}.

The value obtained for the analogous Co–Co bond in $Co_2(CO)_8$ is only 54 kJ mol^{-1}. The low bond strength of this unusual compound is reflected in its behaviour. In the crystalline state there are bridging carbonyl groups between the metal atoms. When the compound is in solution it exists in equilibrium with the form containing the Co–Co bond; and finally, in the gas phase, this latter form predominates. The mass spectrometer has revealed the existence of even more complex metal carbonyls, such as the trinuclear species $Fe_3(CO)_{12}$ and $Os_3(CO)_{12}$, and has shown that the three metal atoms are joined together. The spectrum indicates a stepwise loss of neutral carbon monoxide molecules with the formation of the Fe_3^+ and Os_3^+ ions. The metal cluster may contain more than one element, as in $RuFe_2(CO)_{12}$, and more than three metal atoms, as in $Rh_4(CO)_{12}$ and $Rh_6(CO)_{16}$.

While the transition metal carbonyls have been known for many years, more recently inorganic chemists have made a great variety of organometallic compounds in which the carbonyl groups have been partially or totally replaced by other organic groups. A number of these compounds have vapour pressures high enough to permit their study by mass spectrometry, and their spectra have often been used to determine

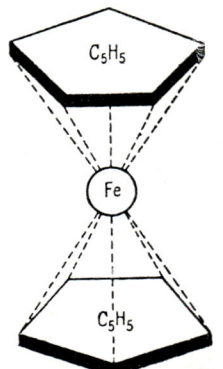

Fig. 3.7. Structure of ferrocene.

their structure. A widely investigated class of compounds is the so-called π-cyclopentadienyl metal derivatives. The first example of these to be discovered was ferrocene, in which the iron atom is symmetrically disposed between the two planar hydrocarbon groups as shown in fig. 3.7. The stability of such a structure is reflected in the mass spectrum, which contains a molecule ion peak of high intensity. This molecule ion eliminates the iron atom, with the simultaneous linking together of the two cyclopentadienyl rings to form the ion $C_{10}H_{10}^{+}$. The spectrum also reveals peaks at much higher mass values, and these have been attributed to ions in which there is a double-decker sandwich involving two iron atoms and three cyclopentadienyl groups. These

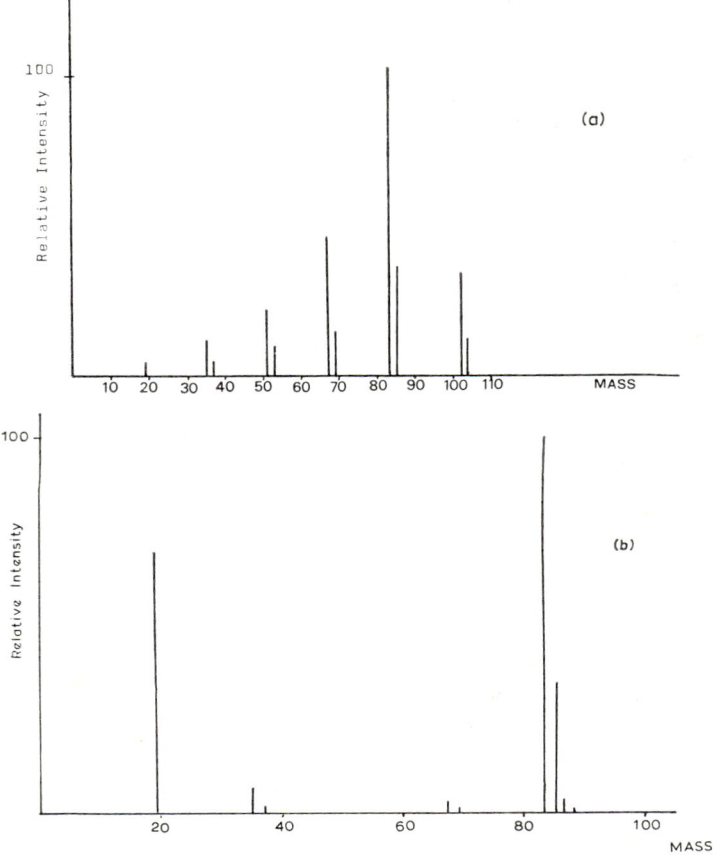

Fig. 3.8. (a) Positive and (b) negative ion spectra of perchloryl fluoride (ClO_3F).

ions may be formed by ion–molecule reactions in the source or may exist in the original lattice.

The iron atom of ferrocene may be replaced by many other metal atoms and one of the cyclopentadienyl groups by a carbonyl or a nitroso group, but the most interesting substitution is when one of the layers is provided by a benzene ring. A number of sandwich compounds have been prepared in which metal atoms are linked to two aromatic rings as in π-dibenzene chromium. In distinguishing such compounds from isomeric species, in which one layer is a cyclopentadienyl ring and the other a tropylium ring, the mass spectrum can be invaluable.

Finally, the rings may be wholly or partially replaced by unsaturated aliphatic compounds such as ethene, as in π-cyclopentadienyl diethene rhodium or π-cyclopentadienyl-butadiene cobalt. Structural identifications are assisted by the stepwise dissociation of the molecule ion, the unsaturated groups being removed as entities as in the case of the metal carbonyls. If, however, the groups around the metal atom contain fluorine rather than hydrogen atoms, the dissociation pattern is complicated by a substantial rearrangement process, in which one or more fluorine atoms are transferred to the metal to form MF^+ or MF_2^+ ions.

3.7. *Negative ions*

When electrons interact with molecules, in addition to processes in which positive ions are formed there occur attachment reactions in which negative ions are formed. In general the cross-sections for these reactions are much smaller than for the corresponding positive ion production. A notable exception to this generalization is provided by halogen compounds, particularly those of group VI elements. The oxygen atom itself has quite a high electron affinity, 1·45 eV, and this is often used as a standard in negative ion studies. An unusual example is provided by perchloryl fluoride, ClO_3F. This is a stable compound with a normal positive ion mass spectrum (fig. 3.8) and is alone in providing a negative ion spectrum in which the ClO_3 ion is the most abundant in either positive or negative spectra. Another interesting compound with a high electron affinity is sulphur(VI) fluoride, and this property together with its inertness and stability has made it important commercially as an insulating gas. It is used in the cases of high voltage equipment to prevent the formation of discharges by trapping electrons and converting them into slow-moving ions. This capture process has been shown to occur over a very restricted range of electron energies, the maximum being at 0·08 eV. The positive ion mass spectrum shows no molecule ion peak, the most favoured process being the production of SF_5^+ and F^-.

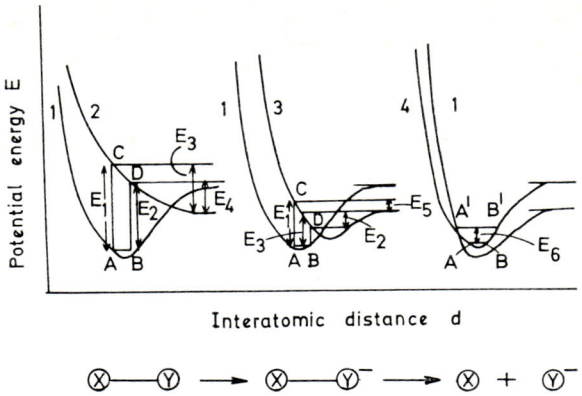

Fig. 3.9. Potential energy curves illustrating the formation of negative ions from diatomic molecules.

Negative ion formation may be understood by reference to the potential energy curve diagram in fig. 3.9. Here the vertical axis represents the potential energy of a diatomic molecule XY and the horizontal axis the interatomic distance. On this diagram curve 1 represents the ground state of the molecule XY. Curves 2, 3 and 4 represent possible states of negative molecular ions. Transitions resulting from the capture of electrons are represented by vertical lines. In the first case illustrated by curve 2, the energy in the negative ion XY⁻ is greater than that required for infinite separation of X and Y⁻, so that dissociation takes place. No stable molecule negative ions can be formed because, unlike positive ion formation, there is no way of losing energy in the form of the translational energy of an ejected electron. In fact the additional energy is lost both by bond-breaking and as the relative kinetic energy of X and Y⁻. In the diagram this will lie between E_3 and E_4. The threshold for the process is at an energy of E_2 and the probability of reaction drops rapidly above an energy E_1. The rate at which this dissociative attachment takes place (known as its cross-section) depends upon the rate at which X and Y⁻ fly apart, and the rate at which the molecule and the electron fly apart to reverse the original reaction. In the second case, illustrated by curve 3, some of the final states have energies below that required for dissociation. Dissociation will only take place if the incident electron energy is between E_1 and E_3. The relative kinetic energy of X and Y⁻ will then lie between zero and E_5. Transitions resulting from the impact of electrons having initial energy between E_2 and E_3 will then result in the formation of vibrationally excited negative molecule ions. In order to survive, these excited ions must lose their energy, and they can only do this by colliding with gas molecules. Thus the concentration of negative molecule ions will be directly proportional to the total gas pressure. In the third case, represented by curve 4, the molecule must be vibrationally excited to the state indicated as A' B' before the attachment reaction can take place. The resulting vibrationally excited molecule negative ion can then be stabilized by collision, or it may dissociate by electron detachment. The incident electron must have sufficient energy for the initial excitation process which precedes the attachment and this is represented in the diagram as E_6.

It can be seen from this simple analysis why it is difficult to get large concentrations of negative molecule ions at the low pressures normally encountered in mass spectrometry. A further feature of negative ion mass spectrometry is that ions are formed within a fairly narrow range of electron energies; that is to say, they are formed by a resonance process. The energetics of the attachment process may be expressed in the following equation:

$$A(Y^-) = D(X-Y) - E(Y) + KE(X, Y^-),$$

that is, the appearance potential for the formation of a negative ion Y^- from a neutral molecule XY, $A(Y^-)$, is equal to the bond dissociation energy $D(X-Y)$ minus the electron affinity of Y, $E(Y)$ plus the kinetic energy of separation of X and Y^-, KE. If the electron affinity is high enough, it will compensate for the energy required for the bond dissociation. The electron affinities of halogen atoms are about 3 eV and comparable with many bond dissociation energies. The result in any case

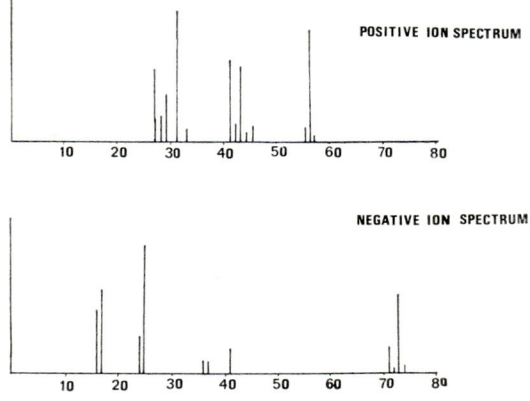

Fig. 3.10. Comparison of positive and negative ion spectra of butan-1-ol (relative molecular mass 74).

is that most attachment processes will take place at comparatively low incident electron energies.

There is a further way in which negative ions may be formed, and this has been called polar dissociation. Here both positive and negative ions are formed, and once again this process is favoured if the molecules contain atoms having high electron affinities. The corresponding equation for the energy required for polar dissociation is:

$$A(X^+) \text{ or } A(Y^-) = D(X-Y) + I(X) - E(Y) + KE(X^+, Y^-)$$

In general, positive ions are formed at lower energies and negative ions at higher energies, by polar dissociation, than by the corresponding ionization and attachment processes. This is not a resonance process and can take place over a wide range of initial electron energies. It is the reason why, in positive ion mass spectrometry, the positive molecule ions of halogenated molecules such as sulphur(VI)fluoride or tetrachloromethane are of such low abundance. Negative ion mass spectrometry may be carried out by making some comparatively minor alterations to the conventional instrument and an example of a negative ion spectrum is shown in fig. 3.10.

3.8. *Halogen compounds*

The study of the very reactive halogens has to be undertaken with caution in a mass spectrometer. It is not unusual to find the spectra containing traces of metal halides, resulting from the action of the halogen on reactive metal surfaces within the machine. An example is the appearance of tin bromide spectra from the action of bromine on soft soldered joints. However, with prior conditioning, the spectra of all the halogens have been recorded, including the mixed molecules such as ICl and BrCl. The spectra of astatine, the halogen missing from the early periodic table and recently produced synthetically, and the mixed halogens AtF, AtCl, AtBr and AtI have been published. Ionization and appearance potentials have been combined to give bond dissociation energies and heats of formation for the halogens, and this procedure has also been applied to the more complex polyhalogen compounds such as BrF_5 and IF_7. Interaction of these species with oxygen has been shown to lead to the formation of compounds such as IOF_5.

3.9. *Compounds of the noble gases*

The mass spectrometer has played an important part in the development of the chemistry of the noble gases. As early as 1936 it was shown that diatomic noble gas ions could be formed by the interaction of an excited and a ground state noble gas atom:

$$He^* + He \rightarrow He_2^+ + e.$$

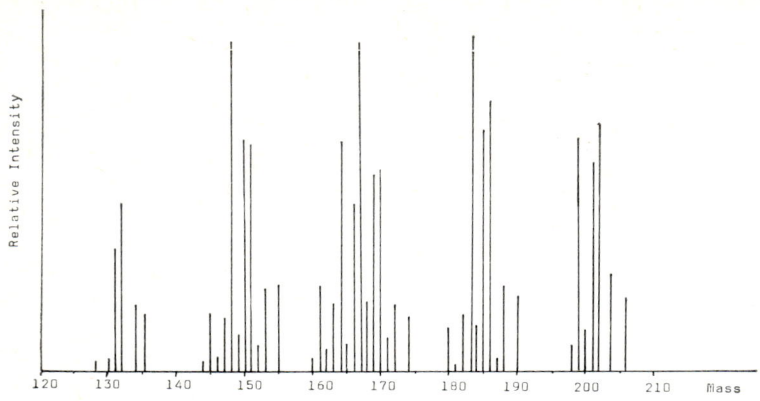

Fig. 3.11. Mass spectrum of a noble gas compound XeO_2F_2.

Later it was shown that at high pressures where appreciable numbers of three-body collisions could occur, samples of noble gases contained a few per cent of diatomic molecules which were not formed in an excitation process. Since 1962 the chemistry of the noble gases has become extensive, and many of the noble gas compounds have been examined mass spectrometrically. The mass spectra of xenon trioxide and tetroxide have been recorded, showing that the bonding is sufficiently strong to survive electron impact. All the fluorides of xenon, XeF_6, XeF_4 and XeF_2, have been studied. Both positive and negative ion spectra give peaks corresponding to ions containing both elements. Appearance potentials of fragment ions have been used to calculate the heats of formation of the molecules and the strength of the bonds, giving results in good agreement with theoretical predictions.

The existence of xenon oxide tetrafluoride was first indicated by the observation of the $XeOF_4^+$ ion in the spectra of impure samples of XeF_4. Early work with xenon hexafluoride showed that $XeOF_4$ resulted from the reaction between the hexafluoride and glass or silica. A second oxyfluoride XeO_2F_2 was also discovered mass spectrometrically. When xenon hexafluoride was allowed to come into contact with water vapour, the mass spectrum contained peaks due to an oxyfluoride. Later pure samples of the compound were prepared by the reaction of xenon trioxide and xenon oxyfluoride ($XeOF_4$). The most complicated noble gas compound which has been examined is xenon tetrakis-trifluoroethanoate, $Xe(CF_3COO)_4$, which is obtained as an unstable yellow solid by the interaction of xenon tetrafluoride and trifluoro-ethanoic acid. The mass spectrum does not contain a peak due to the molecule ion but there are fragment peaks such as XeFCOO$^+$, XeFCO$^+$ and XeCO$^+$ (fig. 3.11). The spectrum is complicated by the variety of naturally occurring isotopes of xenon (see Appendix table).

3.10. *Spark source mass spectrography*

If it was the petroleum industry which stimulated the commercial production of the electron bombardment mass spectrometer, then it was the electronics industry, with its demand for the analysis of high purity semiconductor materials, which was responsible for the production of the spark source mass spectrograph. Many inorganic compounds are very involatile and cannot be evaporated unchanged, even in a Knudsen cell furnace. Even when the inorganic samples can be evaporated and qualitative identifications are possible a semiquantitative analysis cannot be achieved because of the differential rates of evaporation of the components of a mixture. For these reasons there was a demand for an ion source which could produce ion currents representative of the elemental composition of the sample, even if the sample consisted of a mixture of refractory substances. These requirements were met by the invention of the high frequency spark source, which has become a standard component of instruments used in elemental analysis. In its simplest form this consists of a pair of electrodes situated in the vacuum system of the instrument at the entry to the analyser section. The electrodes are connected to a high frequency oscillator operating at a frequency of about 1 MHz with a peak voltage of above 10 kV. The electrodes are quite small, about 20 mm long and a few mm in diameter. They are placed close together, and when the oscillator supply is connected a high frequency spark is developed between the electrodes, which provides a copious supply of ions of the electrode material. Under these conditions the composition of the ion beam is representative of the composition of the electrode material. If the sample is a conductor, the electrodes may be made directly from the sample material, provided that enough is available. If it is not a conductor, then it is most common to mix it in powder form with graphite and compress the mixture into electrodes.

Much of the early work on the spark source was carried out by the pioneer mass spectrometrist, Dempster, but it was not until a suitable mass analyser was designed that the radio frequency spark source found general application. The main problem was that the ions produced in the discharge had a range of velocities and so a velocity focusing analyser was essential. However, once the theory of double focusing had been published by Mattauch and Herzog, it was realized that this system provided a complement to the high frequency spark

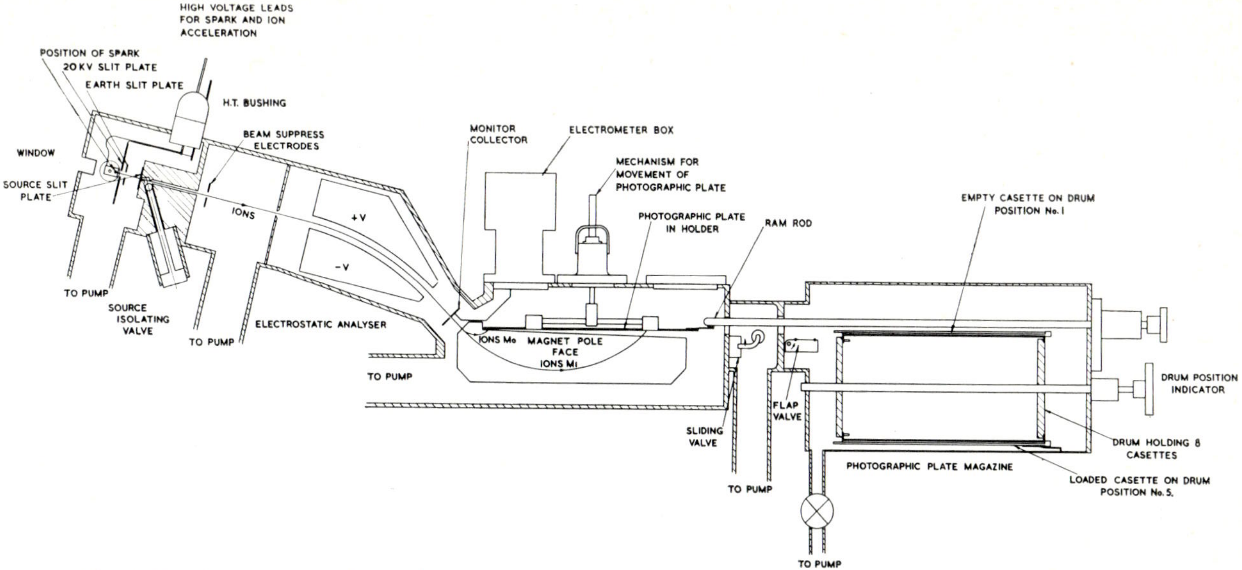

Fig. 3.12. Ion optics of the double-focusing geometry of Mattauch and Herzog MS7 mass spectrograph.

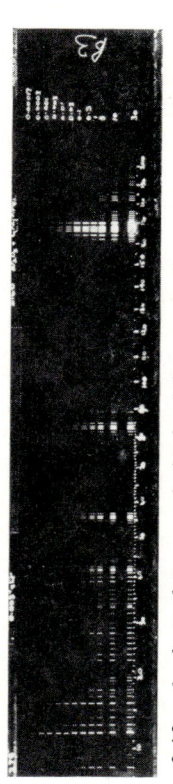

Fig. 3.13. A photoplate record obtained from a spark source mass spectrograph.

source. By 1958 commercial instruments were being designed, a typical example being the MS7 marketed by the A.E.I. Company (fig. 3.12).

There was a further problem associated with the spark source, its instability, the ion current varying with time in a quite irreproducible manner. This was overcome by using the simplest of all integrating detectors, the photographic plate. This was most convenient because, in the double focusing system employed, the deflections due to the electrostatic and magnetic sectors were in opposite directions, and this resulted in the focusing of all ion beams of whatever mass in a single plane (fig. 3.13). Thus a photographic plate situated in this plane recorded all ion beams simultaneously. Each element present was represented by an image of the source slit at an appropriate point along the plate (fig. 3.13).

It was an essential condition for making quantitative measurements that the darkening of the plate should be proportional to the intensity of the ion beam and hence to the amount of the element responsible in the electrode material. Measurements were thus made by densitometry of the plates, and fig. 3.14 shows a microdensitometer trace of an exposed plate. The height of the peaks on the trace is related to the optical density and hence to concentrations in the electrode. Quantitative analysis using the spark source mass spectrograph is complicated by the variations in sensitivity which arise from the use of photographic

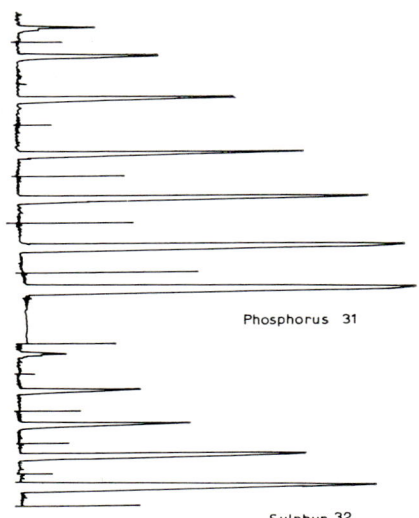

Fig. 3.14. Microdensitometer trace across a photoplate record.

materials and the need for internal standards. Some attempts have been made to eliminate these problems by improving the stability of the spark source and using electrical detection, by measuring the intensity of each ion current as a fraction of the total ion current. However, the simplicity of the photographic plate detector has led to its retention in most modern spark source instruments.

The spark source mass spectrograph can be used to determine the elemental composition of inorganic compounds, although this has not been its common application. The relative amounts of different elements in the sample are usually determined by making a graded series of exposures. The mass lines for the major and minor component element are then selected when they have approximately the same optical density, and the ratio of concentrations is then equal to the inverse ratio of the exposure time needed to produce the lines. It is more common to use the great analytical power of this technique to detect large numbers of elements in very low concentrations simultaneously. This is of the greatest value where ultra-pure materials are required, as in semi-conductor manufacture or superconductor research. Amounts of impurities as low as 1 part in 10^9 have been measured, and this poses problems of calibration. There is a dearth of standard samples and the technique is usually to use a low abundance isotope. Thus one could make a mixture of 1 part of platinum in 10 000 parts of gold and examine the line due to the isotope of platinum ^{190}Pt, which has a relative abundance 0·0127 per cent. This isotope is therefore at the parts per billion level in the original gold sample. One very useful outcome of this ability to measure trace metal impurities is that the distribution of trace elements is frequently characteristic of the place of origin. As an example of its use in archaeology, some small carved stone beads were examined. Not only was the nature of the stone identified but also the area from which it had been mined.

CHAPTER 4
application to physical chemistry

4.1. *General considerations*

IN the 1950's, many of the early production mass spectrometers were used in physical chemistry departments of universities and large research organizations for a very large number of diverse tasks, from studies of combustion processes to the assessment of polymer stability. Two topics in physical chemistry have been selected for discussion. They were chosen quite arbitrarily, and it is not intended to imply that these topics are the most significant or important; they merely provide convenient examples.

The first particular facility which recommended mass spectrometry to the physical chemist was the ability to identify organic and inorganic molecules and measure their concentration in quite complex mixtures. This ability was especially useful in the study of kinetics. It is often necessary to follow the change in concentration of a specific compound with time or temperature, and it is rare that this can be achieved by measuring one bulk property of the reacting mixture, such as pressure or optical density. The mass spectrum of a compound is virtually unique and provides the ready means of identification; further, any changes in the peak heights with variation of other experimental parameters give the change in concentration. For these reasons the mass spectrometer was applied extensively in the study of radical reactions such as those taking place during combustion, photolysis or pyrolysis. An example is the case of propanone (acetone), which when irradiated with u.v. light breaks down into a methyl radical and an ethanoyl (acetyl) radical. These radicals may combine to form ethane and butanedione (diacetyl), or they may abstract hydrogen atoms from the parent ketone to yield methane and other more complex products. The rates at which these secondary processes occur may be followed by measuring the rate of accumulation of products. Then some idea of the energies involved in these secondary processes may be gained by measuring the way in which this rate of accumulation changes with temperature.

There is, however, another way in which propane may be studied using a mass spectrometer. The energy in the incident electrons required to break a chemical bond and ionize the radicals formed may be measured. In other words, appearance potentials can be measured and bond dissociation energies calculated. Thus it is possible to compare energetic information obtained in two entirely different ways with the same machine. In some cases the reaction being studied could be carried out thermally in the instrument itself.

4.2. *The reactions of ions and molecules*

A large number of reactions occurring in the liquid phase, more particularly in aqueous solution, involve charged species. Most students accept that free ions can exist in solutions where the solvent has a high relative permittivity (e.g. water, ammonia, ethanonitrile, glacial ethanoic acid or N,N-dimethylmethanamide); the electrostatic force between ions of unlike charge is greatly reduced by the shielding effect of the solvent molecules which cluster around the ions. The idea that gaseous ions could have an independent existence outside the vacuum system of a mass spectrometer is less familiar, and the investigation of the reactions of such gaseous ions is comparatively new.

It has been known for many years that the sun continually bombards the outer layers of the Earth's atmosphere with charged particles, and that many reactions between ions and molecules, upon which the actual survival of life depends, take place in the thin rarefied air above our heads. Such reactions may influence the concentration of ions which occupy the layers responsible for reflecting radio waves back to Earth. At the same time, many things are going on at ground level which involve the interaction of gaseous ions and molecules. Every time there is an electric discharge some ionization occurs, and our increasing use of highly radioactive substances also increases the amount of ionization caused by their radiation.

The great growth in interest in the effects of ionizing radiation upon matter which took place after 1945, stimulated investigation into the fundamental reactions involved in what came to be called radiation chemistry. As so often happens, the clues had been uncovered many years previously. J. J. Thomson, using his apparatus for the study of positive ions, had noticed a strange phenomenon. The vacuum pumps of the day were, by modern standards, primitive, and it was a common experience to have a blank spectrum which showed evidence for a variety of residual gas molecules. Water adsorbed on the inner surfaces of the apparatus was always difficult to remove, and the water vapour spectrum was only too familiar at mass values 16, 17, and 18. When, however, the concentration of the water vapour rose too high, the water spectrum began to change and a peak at a mass value of 19 became increasingly important. It is very easy to demonstrate this effect with any present day instrument by injecting a large pressure of water vapour into the vacuum system. The intensity of this new peak at mass 19 was proportional to the square of the water vapour pressure, and so had to be the result of an interaction involving two molecules or their ions. The reaction responsible was correctly identified as a hydrogen abstraction reaction in which the molecule ion of water removed a hydrogen atom from a water molecule and formed the hydroxonium ion and a hydroxyl radical:

$$H_2O^+ + H_2O \rightarrow H_3O^+ + OH\cdot$$

In the development of mass spectrometers these ion–molecule reactions were considered to be an irritating complication, and instruments were designed to reduce their contributions to the final mass spectrum to a minimum. When the study of ion–molecule reactions by more conventional means proved difficult, experimenters produced instead instruments deliberately constructed to allow the ion–molecule reaction to predominate.

The methods which had been developed in the past for the study of simple gas reactions proved to be largely ineffective for the study of the reactions which resulted from the passage of ionizing radiation through gases. When purely thermal, or even photochemical, reactions are being studied it is often sufficient to heat or illuminate the gas for a selected period of time and then to analyse the resulting mixture of products. From an identification and measurement of the concentration of all the major products, it is frequently possible to identify the reactions by which they are formed. This is the case even when the reactions involve transient intermediates, such as excited states, atoms or radicals. In the case of radiation chemistry it is rarely possible to do this, because superimposed upon the ground-state chemistry is the chemistry of ion–molecule reactions. In some cases the same compounds are produced by pyrolysis as by radiolysis; thus methane will lead by either process to a complex mixture of higher hydrocarbons. However, the routes by which these are formed are quite different in the two cases.

Returning to the problem of radiolysis at normal pressures, where the ion formed has the opportunity of interacting with a molecule before it loses its charge, it is important to realize that these reactions are very fast. This must be the case, because otherwise the ion first formed would recombine with the electron which had been ejected before further reaction could take place.

There are two simple techniques which can be employed in attempts to differentiate between radical–molecule and ion–molecule reactions in radiolysis systems. The first depends on the suppression of radical reactions by introducing into the system a stable but reactive molecule which has a high affinity for radicals. Thus, in the radiolysis of methane, any reactions due to the intervention of the methyl radical could be suppressed by the addition of either nitrogen oxide (nitric oxide, NO) with which it reacts to form nitrosomethane, or iodine with which it rapidly forms iodomethane. Any ethane produced under these conditions might be fairly considered to be due to ion–molecule reactions. The second idea is to add to the radiolysis system not a reactive substance but a completely inert one; that is to say, a substance with no ground-state chemistry. A few years ago this criterion would have applied to all the noble gases, but today it applies only to the lighter members of the class. Nevertheless, if the addition of small amounts of helium or argon to a radiolysis system results in either a change in rate or product, then it is safe to assume that this change is due to an ion–molecule reaction involving the noble gas. Thus in the case of the radiation-induced exchange of hydrogen and deuterium, some enhancement of the rate takes place in the presence of xenon. The xenon ion could both abstract a hydrogen atom (as in the case of the water molecule ion) and transfer a proton to either a hydrogen or a deuterium molecule. The ambiguities of this type of approach are obvious and it is not, therefore, surprising that the experimenters returned to mass spectrometry as the most convenient tool for the study of the reaction of ions and molecules.

The necessary modifications to the conventional mass spectrometer are comparatively simple. Since the analytical section of the instrument, whatever its nature, must operate at a very low pressure to avoid those very reactions which we wish to study in the source, the two regions are separated by a plate having a very small hole to permit the passage of ions from the source to the analyser. The efficiency of the pumping must be improved to deal with the gas leaking from the high pressure source region through the aperture. Finally, the emission of the filament must be boosted to provide sufficient ionizing electrons in the high pressure gas, and means found for preventing the formation of a gaseous discharge by which the high source potential of some thousands of volts could leak to earth. With such alterations the instrument is capable of giving useful information about the mechanism and rate of ion molecule reactions, provided these are restricted to the interaction of molecule or fragment ions with the original uncharged molecules from which the ions are formed.

The experiment must be conceived as a multi-stage process. In the case of the hydrocarbon methane, the first process is its ionization by electron impact:

$$CH_4 + e \rightarrow CH_4^+ + 2e.$$

The excited methane ion may then dissociate in a variety of ways, each surviving ion being responsible for a mass peak in the resulting mass spectrum. Because of the high probability that any of these ions can encounter a methane molecule during their acceleration to the exit aperture, a series of secondary reactions can take place, the simplest being the interaction of the methane ion with the methane molecule to produce the unlikely looking species CH_5^+.

While the use of a high pressure mass spectrometer is a simple experimental approach to the study of ion–molecule reactions, there are some problems in the interpretation of results. First, it is necessary to determine in the high pressure spectrum which are the peaks due to ions produced by the primary electron bombardment process and which are the peaks due to ions produced by the secondary collision process. In the simplest case, those ions producing peaks which are not visible in the normal low pressure mass spectrum are the result of secondary processes. In more complex cases it may be necessary to plot the variation in peak height for the suspect ion against source pressure. The peak height for a primary ion will be directly proportional to the source pressure, while that for a secondary ion must be proportional to the square of the source pressure. Alternatively one can vary the time spent by an ion in the high pressure source region by increasing the positive voltage on the source electrode known as the ion repeller. At higher voltages the ion spends less time in the source and has less chance of undergoing reaction. There is little effect upon the primary ion spectrum. The simplest way of identifying the reactants or the precursors to any particular ion in a high pressure mass spectrum is to vary the ionizing electron voltage. If, for example, a mixture of argon and hydrogen in a high pressure source is ionized, there will be in the high pressure mass spectrum a peak due to the ion ArH^+, and this fulfils all the criteria for a secondary ion, but how does it arise? There are two possible alternative routes:

$$Ar^+ + H_2 \rightarrow ArH^+ + H$$
$$H_2^+ + Ar \rightarrow ArH^+ + H.$$

The rate of production of ions will be dependent on the electron energy, and this dependence will differ for the two species Ar and H_2 because of their different cross-sections for electrons.

A graph of the ratio of the peak heights due to the primary and secondary ions against electron energy will be a straight line for values of the electron energy above the ionization potential. On the other hand, a graph of the peak height ratio against electron energy for a pair of primary and secondary ions which are unconnected (i.e. the secondary ion does not arise from a reaction involving the primary ion) will be a curve. Applying this idea to the present case, a plot of ArH^+/Ar^+ against electron energy is a straight line, while that for ArH^+/H_2^+ is a curve, showing that the precursor of the secondary ion ArH^+ is Ar^+ and not H_2^+.

The main drawback to this simple method of studying ion–molecule reactions is the uncertainty about the velocity of the ion. The electron impact process creates the ion at rest or with only thermal velocity, and the ion is then accelerated on its way out of the high pressure region. At any time during this acceleration period an encounter with a molecule may take place, and so the high pressure mass spectrum is only an average of all these encounters over a range of ion velocities. Some processing of the data has to take place before any meaningful quantitative findings can be deduced. For these reasons many experimenters have constructed far more elaborate apparatus in which the projectile ion is created in one ion source and isolated, using the mass selection system of one mass spectrometer. Its velocity is then adjusted to the desired value, and it is then allowed to pass into the ion source of a second mass spectrometer where it encounters the appropriate pressure of the target gas. The products of the ion–molecule reaction are then identified. For the charged species it is merely necessary to accelerate them from the collision region into the analyser system of the second mass spectrometer. The unionized products may be identified by using the electron bombardment facilities of the second source. If one of the instruments is sufficiently light, as in the case of a quadrupole mass analyser, it may be mounted on a goniometer and the angular yield of products studied.

The ultimate sophistication is to study the collision process under molecular beam conditions, so that the direction of motion of the gas molecules is accurately defined. With such an arrangement, a beam of atoms or molecules passes into the source, where it intersects the ion beam from the first mass spectrometer. The temperature, and hence the average velocity, of the gas may be controlled by heating the molecular beam furnace. If atoms are being studied, it is even possible to isolate species having particular spin orientation by initially passing the molecular beam through a transverse magnetic field. It is now possible to vary the angle of incidence by adjustment of the relative positions of the molecular beam and the mass spectrometer and still study the angular variation in yield of the products. With such complex equipment some fundamental information about the very nature of chemical reactions may be collected.

With all this equipment, however, some considerable differences will be seen in the way in which ion–molecule reactions are investigated when compared with those normally employed for the study of molecule–molecule reactions.

When a gas reaction is studied it is initiated thermally or photochemically and then the rate of accumulation of products is measured. This rate of appearance is equated to a rate constant multiplied by the concentrations of the reactants. Thus, for example, in the oxidation of hydrocarbons one might study the attack of hydroxyl radicals on carbon monoxide, a process which occurs in the outer zones of most hydrocarbon flames:

$$OH^{\cdot} + CO \rightarrow CO_2 + H^{\cdot}$$

The rate of formation of carbon dioxide would then be equal to a rate constant multiplied by the concentrations of carbon monoxide and hydroxyl:

$$\text{Rate}_{CO_2} = k[CO][OH^{\cdot}]$$

This simple equation expresses the overall sum of all encounters between the two species and in any reasonably-sized population there is a very wide range of relative velocities. To gain some information about the energetics of this elementary reaction the variation of rate (and hence the rate constant) with temperature must be studied. From simple gas kinetic collision theory k may be expanded in terms of A, the pre-exponential factor and E the activation energy, which are incorporated in the Arrhenius equation:

$$k = A \exp(-E/RT)$$

where R is the gas constant and T is the absolute temperature. The exponential part of the equation expresses the fact that at any given temperature the reactants have a wide distribution of velocities. To find values of E it is necessary to plot the logarithm of the rate constant against $1/T$. From the slope of the graph, E may be calculated; and from the intercept, A may be determined.

The same considerations cannot be applied to the case of ion–molecule reactions. The ions do not have a natural distribution of velocities, and they all tend to have the same direction of approach. It is, therefore, more convenient to treat the ion–molecule system from the physicist's rather than the chemist's point of view, and calculate a cross-section rather than a rate constant for the reaction. It can be seen that the situation is formally analogous to the absorption of light by matter; and that the rate of attenuation of an ion beam, on passing into a chamber of target gas, is proportional to the depth to which the beam has penetrated (x) and to the concentration of the molecules which it encounters along its path (N). Thus:

$$-dI/I \propto N dx.$$

If the proportionality constant q is the area which the target molecule presents to the projectile ion, then it is analogous to the molar absorptivity for optical absorption. By integration the equation becomes:

$$I = I_0 \exp(-Nqx).$$

This is of the form of Beer's Law and has similar restrictions. It is only true for one form of interaction and for a precisely defined relative velocity of ions and molecules. If this relative velocity is v, the relationship between the rate constant and the cross-section q is:

$$q = k \times v.$$

It can be seen how approximate values of q may be determined from the high pressure mass spectrometer experiment from simple measurements of the primary and secondary ion currents. A typical value for q determined in this way is 10^{-15} cm^2.

The molecules present a large collision cross-section to the approaching ions; much larger than they do to an approaching molecule or radical. A simple theory to explain this behaviour was first put forward by Langevin and, although it has been superseded by more advanced treatments, it gives at least a qualitative picture of the situation. An ion, having a charge e and being at a distance r from the molecule, polarizes it, that is to say, it produces a distortion of the outer electron shells so as to create a dipole $\alpha e/r^2$ where α is the polarizability of the molecule. This dipole is then attracted by the ionic charge, the force between the ion and the dipole being $\alpha e^2/r^4$. This force rises rapidly as r decreases until the approach is so close that repulsive forces come into play and the ion begins to orbit around the molecule. There is thus a long interaction path and a high cross-section. A simple relationship has been derived to express the cross-section in terms of the polarizability and the relative velocity:

$$q = \frac{2\pi}{v}\left(\frac{\alpha e^2}{m}\right)^{1/2}.$$

In cases where this can be checked with any certainty, there is a fair agreement.

The most interesting feature of ion–molecule reactions is the variation of q with v, or the accelerating voltage V applied to the projectile ion. For voltages down to 50 V the most efficient reaction is that of charge exchange, and the highest values of q are obtained for the symmetrical or resonance situation, in which the ion and molecule are the same chemical species. This has some interesting results, since it provides a mechanism by which the energy of electromagnetic radiation may be transferred to a gas containing charged particles. The ion moves under the influence of the oscillating potential gradient and then strikes a molecule. Charge is exchanged, but no momentum, so the ion is now at rest and the uncharged molecule possesses kinetic energy. The potential gradient may now be reversed and the rest ion accelerated to lose energy again by a further collision. This process is responsible for the absorption of radiofrequency radiation by flames.

When the ion and the molecule are not the same chemical species, the ionization potentials will be different and the reaction may become endothermic; the difference in the ionization potentials must be made up by the conversion of the kinetic energy of the ion into electronic energy. For such a reaction the value of q will rise rapidly from a threshold. An example of such a reaction is the ionization of nitrogen by fast xenon ions.

When the target is a molecule with a lower ionization potential than that of the projectile, or when the projectile has considerable kinetic energy, then the transfer of charge may be followed by dissociation of the ion. This is called dissociative charge transfer, and the spectra obtained are analogous to those produced by electron impact, although they are not the same. An example of a dissociative charge transfer is:

$$Ar^+ + CH_4 \rightarrow Ar + CH_3^+ + H^\cdot$$

The cross-section shows a similar behaviour rising from a threshold, but this threshold may differ for the different dissociation processes. At ion accelerating voltages below 50 V, charge transfer processes compete with other ion–molecule reactions, the most important being that of atom transfer. The simplest of these reactions has already been described, and a large number of them have been studied. Some typical examples are:

$$N^+ + O_2 \rightarrow NO^+ + O^\cdot$$
$$O^+ + CO_2 \rightarrow O_2^+ + CO.$$

A typical value for q for a hydrogen transfer reaction is 10^{-14} cm^2.

Finally, there are the reactions of ion transfer, the most interesting being those of proton transfer. Here a complex ion transfers both its charge and a hydrogen atom at the same time, as in the case of methane:

$$CH_4^+ + CH_4 \rightarrow CH_5^+ + CH_3^\cdot$$
$$CH_5^+ + C_6H_6 \rightarrow C_6H_7^+ + CH_4.$$

The proton transfer reaction has been exploited in an alternative method of ionization called chemical ionization. Using this technique a table of proton affinities can be produced analogous to electron affinities.

It has only been possible to give a brief mention of this strange chemistry of ions and molecules which occurs in mass spectrometers, in radiolysis and in the upper atmosphere. It is still being actively studied in laboratories throughout the world.

4.3. The study of flames
4.3.1. *Principles*

A flame is a remarkable phenomenon; it is a chemical reaction occurring in the gas phase in a vessel without walls. Further, the progress of the reaction with time is spread out along the direction of propagation of the flame, the flame reactions being initiated at the base and being complete at the edge of the luminous cone. It is thus possible to introduce into the flame a variety of probes which are sufficiently small to prevent any interference with the flame chemistry, but which can provide information about the types of reaction which are going on in a flame and to make measurements of reaction rates.

The probe may be a simple thermocouple to measure the temperature profile, it may be a small Pitot tube used to estimate gas velocity, or more appropriately to the subject of this book, it may be the inlet of a mass spectrometer. It is possible to obtain more information about flame chemistry with this instrument than with any other. Using the most simple apparatus it is possible to identify all the stable species existing in the flame gases. Then by calibration the concentration profiles of all these species can be determined. A more complicated inlet system can allow the unstable species, such as free radicals and atoms generated by the flame reactions, to be withdrawn from the hot gases into the mass spectrometer. In this way they too may be identified and their concentration profiles measured. A further modification of the instrument may be made so that all the uncharged species are rejected and only those species carrying a charge, such as positive polyatomic ions, may be sampled and identified. Finally, the instrument can be provided with a facility which permits the simultaneous measurement of concentration and temperature.

In spite of the familiar sight of the Bunsen flame in the chemical laboratory, there are few who recognize the physical and chemical factors associated with its stability and safety. Flames are of two kinds, stationary and propagating. A propagating flame, in which the flame front moves through the stationary gas mixture, is difficult to control and is frequently terminated by explosions or detonations. The most appropriate vessel in which to conduct such studies is a soap bubble. The stationary flames are those which sit on burner mouths, where the rate of advance of the flame front is balanced by the movement of the unburnt gas through the burner mouth. Stationary flames are of two kinds, pre-mixed and diffusion flames. In diffusion flames the fuel alone passes out of the burner mouth, and combustion takes place with air diffusing in from the surrounding atmosphere. A flame of this type may be exemplified by the Bunsen burner operating with the air hole closed. These flames are rarely steady and are difficult to control. The most suitable flames to examine are of the pre-mixed type, in which the fuel and oxidant are mixed and pass into the burner mouth together. This type is exemplified by the Bunsen burner operating with the air hole fully open. Further discussion will be limited to flames of this type.

Although everyone can describe a flame and recognize it, it is difficult to make a precise definition. The phenomenon is usually associated with the emission of light, but there are flames such as the hydrogen/air flame which emit no radiation in the visible region of the electromagnetic spectrum, while many chemical reactions which emit radiation in the visible region are certainly not flames, e.g. chemiluminescent reactions in solution. A flame is normally considered as a high-temperature process and is usually at atmospheric pressure; but if the pressure is reduced, hydrocarbon oxidation flames may be run at temperatures between 200–400°C, while the oxidation of phosphorus vapour, which is certainly a flame reaction, may occur at room temperature. Finally, a flame is often considered to be an oxidation process, but there are many decomposition reactions which may be run as flames; for example, the thermal breakdown of nitric oxide into nitrogen and oxygen, or the pyrolysis of perchloric acid to form water, hydrogen chloride and various oxides of chlorine.

Most pre-mixed flames exhibit three distinct regions. First, a pre-heat region, in which no chemical reaction takes place, but in which the temperature of the gas rises owing to the diffusion of both thermally energized molecules and active chemical species such as free radicals. Immediately above this zone is the flame front, which is characterized by strong light emission in the case of hydrocarbon oxidation. In the Bunsen flame this is represented by the greenish-blue inner cone. In this region there is a very steep temperature rise and very energetic exothermic reactions take place. In some very specialized flames, such as hydrocarbon–perchloric acid flames, there are two separate flame fronts corresponding to the onset of two different oxidation reactions. Above the flame front there is an outer diffusion sheath, in which any species produced by incomplete combustion in the flame front are burnt (in the case of hydrocarbon flames) to carbon dioxide and water with the aid of extra oxygen diffusing in from the surrounding atmosphere.

If this diffusion is prevented by, for example, surrounding the flame with nitrogen, the flame lifts off the burner. The outer diffusion sheath thus stabilizes the flame, and this is particularly important at the burner mouth where the fall in velocity of the unburnt gases leaving the mouth is balanced by the rise in the flame velocity due to the diffusion in of extra oxygen. If the flame is run at reduced pressure, the boundaries between the two regions become less distinct and finally vanish, so that at low pressure there is little difference between pre-mixed and diffusion flames.

4.3.2. *Burners*

A simple device like a Bunsen burner is inadequate for making a precise investigation into the nature of flames, and a much more complex apparatus is required, designed to operate a steady flame and eliminate all fluctuations in the experimental parameters. An essential part of this equipment is a device for producing a stable flow of mixed gases of known composition. The gas composition is usually adjusted with the aid of needle valves interposed between the gas supply cylinders and a mixing vessel. The gas composition itself may be experimentally determined by applying the sampling probe to the pre-heat zone of the flame, or it may be estimated from readings of flow meters placed in the gas supply lines. In any case the gas flow must be maintained laminar; in other words the gas must be considered to move as a series of parallel layers, the innermost layer moving with the greatest velocity and that closest to the tube wall with the least. With truly laminar flow there is a parabolic distribution of gas velocity across the diameter of the tube, and there is no component of gas velocity at right angles to the direction of flow. Departure from this state leads to turbulence in the gas flow and ultimately in the flame itself. Laminar flow is best maintained by having a long straight approach tube to the burner mouth.

Burners may be of two types, open or screened. The first type is similar to the familiar Bunsen, and the actual shape of the flame is determined by the distribution of gas velocities across the diameter of the mouth. Under normal conditions, the cross-section of the flame front would approximate to a parabola, but if a nozzle is added then the cross-section may become triangular. Other disturbances to the gas flow by the introduction of baffles or grids may create more extreme modifications of flame-front geometry, as in the Meker burner. More control of flame conditions can be achieved with the second class of burners, in which the mouth is closed, with a porous screen through which the gas mixture diffuses. The shape of the flame is now completely dependent, not on gas flow, but on the shape of the screen. It is common to run slow-burning flames as a flat button over a horizontal screen, but more recently a novel design has been introduced. The screen is in the form of a hollow sphere and the gas diffuses through the porous screen, emerging along radial lines. The resulting front is also in the shape of a sphere completely surrounding the screen, except for a small area where the inlet tube meets the screen. This type of burner is particularly useful when quantitative studies of flame structure are being made, and it provides a very stable flame.

Whatever type of burner is being used, however, it must be operated in a housing which protects the flame from variations in the ambient conditions. Such a housing must be a pressure vessel, since it is often advantageous to run flames at low pressures. The pressure within the vessel must be kept constant by forming constrictions in the inlet and outlet gas flow tubes. This ensures 'choking' or sonic flow in the constricted regions, which prevents the transmission of oscillations in pressure. Even so, it may be necessary to increase long-term pressure stability by attaching to the burner housing a large ballast volume, maintained at the desired pressure and connected to the burner chamber through a valve, which may be operated only when there is a pressure differential across it. The probes which are used to study the flame characteristics are introduced into the burner chamber through pressure-tight seals in the burner housing.

4.3.3. *Radicals and ions*

Reactions between molecules and molecules have a very high activation energy, since before reaction can occur chemical bonds must be broken. For the reaction:

$$H_2 + I_2 \rightarrow 2HI$$

the energy of activation is 168 kJ mol^{-1}. In the reactions of radicals or atoms with molecules, a chemical bond is formed at the same time as one is broken, so that a much lower activation energy is required. In the following reaction:

$$OH\cdot + C_2H_6 \rightarrow C_2H_5\cdot + H_2O$$

commonly encountered in flames, the energy of activation is only 25·2 kJ mol^{-1}. It might be considered, following this comparison, that radical–radical reactions would preponderate in flames, because the energy of activation for their combination is close to zero. However, this is not the case, because of another factor known as the third body effect. For any two radicals to combine and form a stable molecule, the energy released in forming the new bond must be removed, or the two radicals will be immediately reformed because the energy released on formation of the bond is equal to the energy required to break it. The stabilization of the molecule can only be achieved by the participation of a third body or molecule, which carries away the energy of the newly formed bond as either translational or vibrational energy. Such a result, however, requires an almost simultaneous collision of three particles, which is a comparatively rare event. For this reason the radicals disappear by combination at a relatively slow rate and the flame reactions are substantially radical–molecule reactions. For all these reasons substantial concentrations of free radicals and atoms build up within the flame front. The most striking way of demonstrating the existence of these radicals is to examine the light emitted by flames. In all cases the species responsible for this emission are found to be radicals. Thus, the blue-green colour of the inner cone of the Bunsen flame is due to simple diatomic radicals such as CH, C_2 or CN, while the strong emission in the ultraviolet given by hydrogen–oxygen flames comes from highly excited OH· radicals.

4.3.4. *Scavenging*

Although the existence of large concentrations of free radicals may be demonstrated in this way, accurate measurements of their concentrations have to be made using a mass spectrometer. The problem which has to be overcome is the disappearance of the radicals after sampling, by recombination either in the gas phase or on the walls of the sampling probe. One very simple answer to this problem is to reduce the importance of the recombination process by introducing an efficient competing reaction. Such a reaction, known as a radical scavenging reaction, has a very low activation energy, and its rate is further promoted by providing a relatively high concentration of the scavenger. Suitable molecules to use as radical scavengers are iodine and nitrogen dioxide. Any hydrocarbon radical will react readily with molecular iodine to yield the corresponding iodoalkane:

$$CH_3· + I_2 \rightarrow CH_3I + I·$$

The mass spectrum of iodomethane is easily identified and indicates the presence of the methyl radical in the reaction mixture. Assuming that the reaction is quantitative, it is possible to measure the concentration of methyl radicals by comparing peak heights in the spectrum with those obtained with known concentrations of pure iodomethane. This

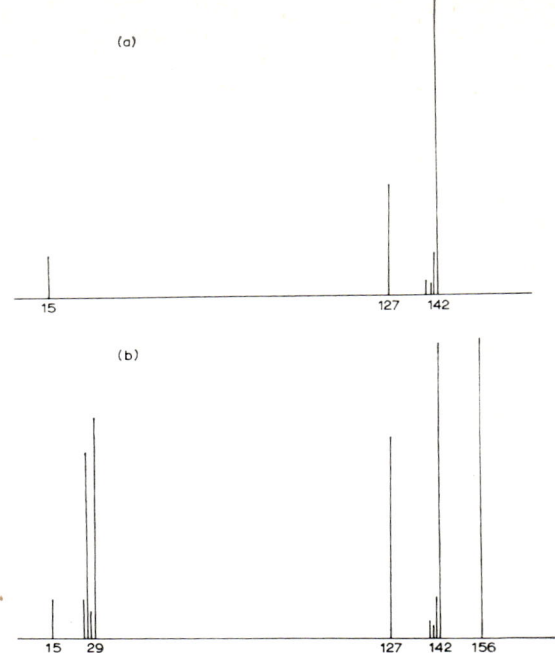

Fig. 4.1. Mass spectrum of (a) iodomethane, (b) mixture of iodomethane and iodoethane.

technique may be extended to flames in which more than one hydrocarbon radical occurs. The mass spectrum shown above is of a mixture of iodoalkanes obtained in this way (fig. 4.1). Each iodoalkane, and consequently each radical, may be separately identified and determined.

The reactions of nitrogen oxide with hydrocarbon radicals are more complex, and it is not a suitable scavenger for large radicals. Nitrogen dioxide, however, may be used in the determination of some atomic species occurring in flames, e.g. oxygen atoms:

$$O^{\cdot} + NO_2 \rightarrow NO + O_2.$$

The measured concentration of nitrogen oxide has been shown to be exactly equivalent to the oxygen atom concentration before the addition of NO_2.

4.3.5. *Study by mass spectrometry*

In the case of the mass spectrometric analysis of flame gases, the probe consists of a very fine quartz capillary with an internal diameter of a few micrometres. The most satisfactory mode of sampling is to draw the gases into the probe at the same speed as they are moving through the flame, i.e. isokinetically. This can be achieved by exhausting the probe interior with a vacuum pump. The amount of material removed from the flame in this way does not amount to more than a few micrograms per second, but this is sufficient for a complete chemical analysis. The sample may be taken intermittently or continuously, depending on which type of spectrometer is available. In a batch sampling procedure, the gases flowing into the probe are trapped by cooling in liquid air in a removable vessel which is then attached at some later time to the conventional inlet system of the spectrometer. Perfectly satisfactory concentration profiles may be constructed by batch sampling along the axis of the flame and analysing the samples collected using a routine instrument. Where possible, however, it is more satisfactory to sample continuously and to connect the sampling probe directly to the inlet system or the ion source of a mass spectrometer used only for this purpose. It may be necessary to match the pumping rate of the probe and mass spectrometer to keep within the operating pressure range of the instrument. As the molecular species existing in flames will not normally contain very large numbers of atoms, a low resolution instrument is sufficient for this purpose. Nor is it necessary to have rapid scan facilities, because the conditions should remain constant at any one point in the flame as long as sampling continues. The ability to record a mass spectrum in five to ten minutes is all that is required, and the mass resolution need not exceed two hundred. A quadrupole spectrometer is thus ideally suited for flame studies because of its compactness and ease of operation.

The concentration profile is determined by moving the probe through the flame along a direction perpendicular to the flame front and recording a spectrum at each point (fig. 4.2). Such profiles, however, refer only to the stable species existing in the flame gases. It may be anticipated that the very rapid cooling experienced by the sample on entering the probe, due to its expansion into a region of low pressure, would rapidly freeze the reactions and provide a representative sample; but if the gases contain reactive species, such as free radicals or atoms, these will disappear by combination on the walls of the probe. Most flames operate at temperatures above 1000°C in the flame front, and at such temperatures the most thermodynamically stable species may not be molecules but free radicals. There is, for example, a high concentration of hydroxyl radicals in the flame of rocket exhausts; these very rapidly disappear as the gases come to equilibrium with their surroundings. Thus, in simple probe sampling, whether batch or

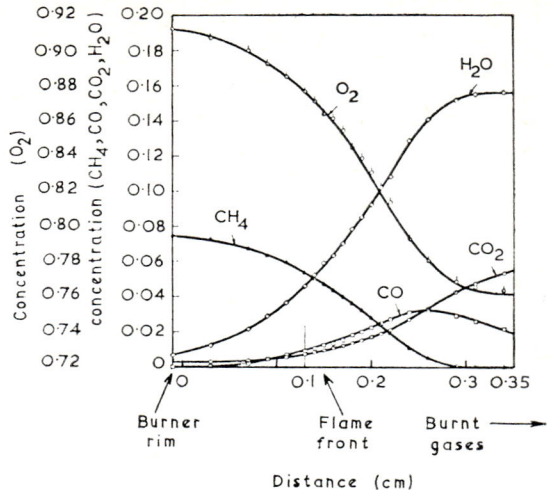

Fig. 4.2. Concentration profiles of stable species in a methane/oxygen flame.

continuous, the presence of free atoms or radicals is not revealed and their contribution is ignored in the final analysis. This is unfortunate, because all flame reactions are substantially radical–molecule reactions.

The apparatus required for the operation of the scavenger technique is extremely simple. The normal fine sampling probe is merely fitted with a side arm through which the scavenger molecules are allowed to pass into the straight section of the probe, where they react with the sample withdrawn from the flame. The reaction products then pass down the tube and are admitted to the ion source of the spectrometer.

Although the simplicity of this technique recommends itself to the flame chemist, it is obviously not possible to make a complete analysis of all radical species existing in a flame in this way. To achieve such an analysis, a more complex inlet system has to be employed. The problem is to remove a representative sample of gases from the flame at a relatively high pressure and transfer it without change to the ion source of the mass spectrometer, which is operating under conditions of high vacuum. The only solution to this problem is 'line of sight' sampling. In this method the molecules move in straight lines from the flame into the ionizing electron beam, without striking any of the walls of the probe or spectrometer. The apparatus used is called a molecular beam system.

The aperture which withdraws the sample directly from the flame is larger than that of the conventional probe, and immediately opposite to it is a second aperture. The space between these two apertures is pumped, using a high pumping speed, to reduce the number of collisions in the interspace. All species with a component of velocity at right angles to the line joining the two apertures collide with the walls and are pumped away. There is, in consequence, a reduction in pressure from the flame to the interspace. Molecular and radical species which pass through the second aperture now enter a second interspace, which they cross on the way to a third aperture. This second interspace is also pumped to maintain it at a low pressure. As there are far fewer collisions in this second interspace the pressure is very much lower. Beyond the third aperture is the ion source, and there is direct access to the electron beam.

In this way only those species which have travelled in a straight line from the flame to the ion source are ionized and detected by the mass

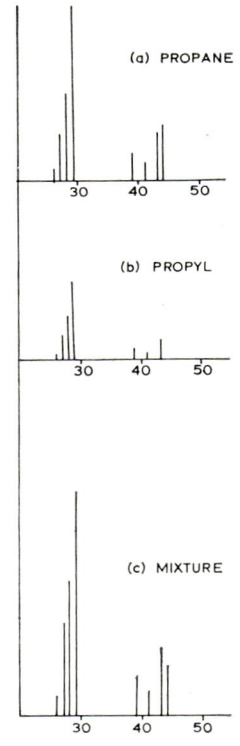

Fig. 4.3. Mass spectrum of (*a*) propane, (*b*) propyl radicals, (*c*) gas mixture containing both propane molecules and propyl radicals. (Computer realization).

spectrometer. The number of molecules or radical species per cm³ in the beam entering the ionization region is not large, and high instrumental sensitivity has to be used. The detection problem is further complicated by the background pressure in the ion source, which may approach that of the beam itself. In order to discriminate against this background it is necessary to interpose between the last two apertures a vibrating shutter, which serves to modulate the beam signals but has no effect upon the background. Thus only those ion currents which have an a.c. component arise from the molecular beam. This discrimination may be carried out electronically in the recording facilities of the instrument.

Even when it is possible to detect the constituents of a molecular beam, however, it may still be difficult to identify the radical species. In order to understand this, let us consider a flame in which propane is being oxidized, and assume that this oxidation takes place with the intervention of a propyl radical. The molecular beam will then contain, among other entities, a large concentration of propane and a much smaller concentration of propyl radicals. The mass spectra illustrated are those of propane, propyl, and a mixture of the two (fig. 4.3). All the mass peaks in the propyl spectrum are present in the spectrum of propane. If a normal analysis of the beam is carried out the only criterion of the existence of propyl would be a slight enhancement of peaks at mass values below 44, compared with the peak at 44 itself. Such a measurement would be very imprecise and an alternative approach is required.

For reasons of sensitivity, a mass spectrometer is operated routinely with an ionizing electron energy of 50 to 70 V. The energy transmitted to the target species by the incident electron is sufficient to result in the formation of a highly excited ion. The ion retains enough energy to enable it to undergo several parallel or consecutive unimolecular decompositions. Some of these are rearrangement processes, but many are simple bond-breaking processes, as in the formation of the propyl ion from the propane ion.

$$C_3H_8^+ \rightarrow C_3H_7^+ + H\cdot$$

If the energy in the ionizing electron beam is reduced, it may be insufficient to enable bond-breaking decompositions to occur. Ideally, the voltage of the electron beam should be reduced to a value sufficient only for ionization. With such a voltage, propane molecules are converted only to propane ions and propyl radicals to propyl ions. The propane spectrum obtained under these conditions is simulated in

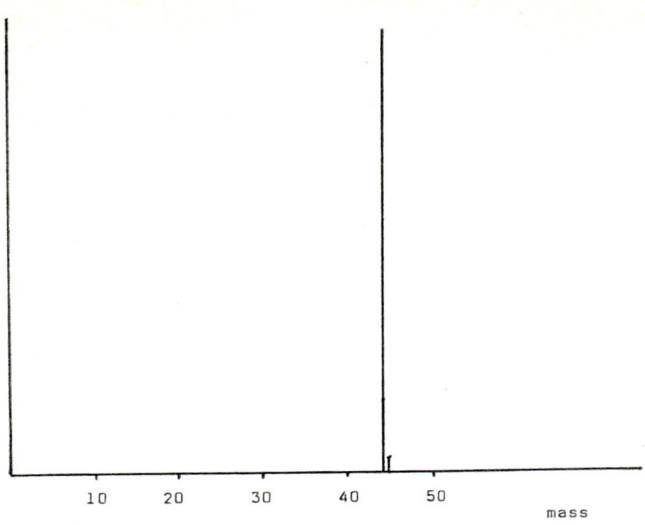

Fig. 4.4. The mass spectrum of propane at a low electron accelerating voltage (12 V).

fig. 4.4. In such a spectrum the peak heights at the mass values 44 and 43 would be proportional to the concentrations of propane and propyl alone. Because of the possible wide differences in the ionization potentials of molecules and the radicals derived from them it may not always be possible to select the ideal voltage. The first ionization potentials of propane and propyl are 11·2 eV (1080 kJ mol^{-1}) and 8·7 eV (840 kJ mol^{-1}), so that an appropriate ionizing electron voltage would be 11·5 eV. It is therefore possible to distinguish between the signals due to molecules and radicals. This facility is only gained, however, at the cost of a further loss in instrumental sensitivity.

The final difficulty is one of calibration, and can be overcome only in certain cases. Initial calibration of the instrument with a pure standard sample is required. A known pressure of a pure sample of the substance to be analysed is introduced into the inlet reservoir of the spectrometer. The substance is allowed to leak slowly into the ion source through a capillary or ceramic sinter. It is essential that the pressure in the inlet reservoir remains constant during the time required to record the mass spectrum or measure the ion currents at selected mass values. It is then possible to make a measurement at a mass value which is characteristic of the sample, and relate the ion current measured at this value to the pressure of the sample in the inlet reservoir by a factor which expresses the sensitivity of the instrument for that particular molecule.

It is obviously not possible to carry out this simple calibration procedure for free radicals and atoms because, with the exception of some unusual species such as NF_2, it is not possible to maintain a known stable pressure of free radicals or atoms at accessible laboratory temperatures and pressures. Instead an alternative procedure, which is less direct but still perfectly valid, must be employed. A known pressure of the dimer of the radical or atom is introduced into the inlet reservoir and the instrument calibrated for this dimer. The dimer is then allowed to flow over a heated surface or pass through a radio-frequency discharge after it leaves the reservoir and before it reaches the ion source. In this region some of the dimer is decomposed into atoms or radicals and, as a result, the ion current due to the dimer molecule ion is reduced. At the same time the ion currents at lower mass values are enhanced. It is now possible to disentangle the spectra due to the dimer and the radical or atom, and to calculate the sensitivity of the instrument for the latter species from the reduction in the ion current due to the dimer molecule ion.

Consider the very simple case of the calibration for the oxygen atom. First the spectrum of the oxygen molecule is recorded, and the ion currents at mass 32 and 16 are measured. The stream of oxygen entering the instrument is then exposed to a radio-frequency discharge and the spectrum again recorded. The ion currents at mass 32 and 16 are once again measured. If the partial pressure of oxygen molecules under the two different inlet conditions are represented by p_1 and p_2 respectively, then the partial pressure of oxygen atoms arising as a result of the radio-frequency discharge is $2(p_1-p_2)$, since every oxygen molecule disappearing will give rise to two oxygen atoms. Representing the measured values of the ion currents under the two inlet conditions as I_a and I_b respectively, the ion current I_c at mass 16 due to the oxygen atoms alone can be calculated.

$$I_c = I_b(16) - \frac{I_b(32) \times I_a(16)}{I_a(32)}$$

The total sensitivity of the instrument for oxygen atoms is:

$$S = I_c/2(p_1-p_2).$$

It is now possible to trace the concentration profile of oxygen atoms through a flame.

One of the most fascinating phenomena associated with flames is the formation of detectable concentrations of free electrons and ions. The origin of charged particles in flames has only recently been investigated in detail. Much of the impetus for this research has come from two directions. In the generation of electricity an attractive increase in efficiency could be achieved if, in addition to the current produced by conventional turbines and alternators, an extra current could be drawn directly from the flame itself. Efforts have been made, therefore, to enhance the concentration of charged particles in the flame. Efforts in the opposite direction, to reduce the charged particle concentration, have been made by rocket engineers in order to reduce interference with radio signals (and to make detection by enemy radar more difficult).

The origin of the separation of charge in a flame is intriguing. At high temperatures there must be a certain amount of purely thermal ionization, and it has been shown that equilibrium exists under these circumstances between the molecule or atom and the ion and emitted electron:

$$M \rightarrow M^+ + e$$
$$K = [M^+][e]/[M]$$

The equilibrium constant, K, in this case is temperature-dependent, and the concentrations of free electrons and ions may be estimated from the Boltzmann distribution $\exp(-E/kT)$, where E is the energy required to ionize the molecule or atom M, k is the Boltzmann constant, and T the absolute temperature. E is normally about 10 eV or 966 kJ mol^{-1}. From this it can be inferred that even at the highest flame temperature the amount of thermal ionization would be very small indeed.

A clue to the paradox is given by the fact that one of the hotter flames, the hydrogen–oxygen flame, exhibits much less ionization than cooler hydrocarbon–air flames. It is clear that this ionization must be a chemical phenomenon, and it has been given the name of 'chemi-ionization'. To understand it one must compare it with another process going on in hydrocarbon flames, chemiluminescence. The greenish-blue light emitted by the inner cones of most hydrocarbon flames is due to the presence of diatomic radicals such as C_2, CN and CH. They have been formed in flame reactions in an excited state, so they are at a much higher temperature than the surrounding gases. They become stabilized by losing this excess energy in the form of a photon. In a similar way other radical species in an excited state may be stabilized by the emission of an electron. For a process requiring this amount of energy to occur, the energy must be released by the formation of chemical bonds. The most likely event would, therefore, be the formation of an ion from a radical with a low ionization potential by the formation of a highly energetic chemical bond as, for example, a multiple bond.

A reaction which exemplifies this reasoning is the combination of oxygen atoms with methyne radicals:

$$CH^{\cdot} + O^{\cdot} \rightarrow CHO^+ + e.$$

The CHO^+ ion has been detected in a variety of hydrocarbon flames and the energetics of its formation are well understood.

The heat of reaction ΔH_f for the formation of the CHO radical can be calculated using thermodynamic data. The carbon–hydrogen bond dissociation energy is known from the absorption spectrum: $D(C-H) = 336$ kJ mol^{-1}. The heat of formation for the CH$^{\cdot}$ radical is obtained by combining this value with the heats of formation of the gaseous carbon atom, $\Delta H_f = 722$ kJ mol^{-1}, and of the hydrogen atom $\Delta H_f = 219$ kJ mol^{-1}:

$$D(C-H) = \Delta H_f(C)_{gas} + \Delta H_f(H) - \Delta H_f(CH) = 605 \text{ kJ mol}^{-1}.$$

This value is then combined with the known heats of formation of the CHO$^{\cdot}$ radical (8·4 kJ mol^{-1}) and the oxygen atom (248 kJ mol^{-1}):

$$D(CH-O) = \Delta H_f(CH) + \Delta H_f(O) - \Delta H_f(CHO).$$

The bond dissociation energy is thus calculated to be 869 kJ mol^{-1}, and this will be the energy released when the CHO$^{\cdot}$ radical is produced by the combination of the methyne radical and the oxygen atom. But the ionization potential of the CHO$^{\cdot}$ radical is 9·3 eV (or 899 kJ mol^{-1}). Thus at the temperature of the flame many of the CHO$^{\cdot}$ radicals produced are stabilized by ionization with the production of a free electron and a CHO$^+$ ion.

It is now apparent why the cooler hydrocarbon flames have a larger concentration of free electrons and ions. In the hydrogen–oxygen flame there are no radicals of sufficiently low ionization potential. This fact is made use of in the widely used flame ionization detector for gas chromatographic studies. This works on the principle that only a very small current can be drawn from a hydrogen–oxygen flame until a trace of carbon-containing compound is added. Then the current rises steeply due to the formation of ions and free electrons. If sensitive electrometers are available to measure this change in current, very small traces of organic compounds (of the order of 1 ng) in the effluents from gas chromatographic columns can be detected.

The current may be withdrawn from the flame quite simply by inserting a pair of electrodes at the base and at the tip and maintaining a potential difference of about 100 V between them. With quite small flames containing a few per cent of hydrocarbon in the flame gases the current can rise to the μA region. The flame, however, does not obey

Ohm's law, and the relationship between voltage and current may be expressed in a quadratic form such as:

$$E = \frac{aI}{d} + bI^2,$$

where a and b are constants related to the specific resistance of the flame gases and the electrode surface, and d is the distance separating the electrodes.

The reason for this behaviour is the difference in mobility between the ions and electrons, a difference which is not due to the disparity in their mass. This difference in mobility can be demonstrated by plotting the fall of potential between the two electrodes with a probe, or by imposing a transverse magnetic field, when opposite sides of the flame become oppositely charged. This is an extension of the Hall effect commonly encountered in semiconductors. The difference in mobility, which is about one thousandfold, is due to the phenomenon of charge transfer. The ion, on acquiring a translational energy from the potential gradient, collides with a neighbouring uncharged molecule. The molecule acquires the electric charge as a result of this collision, but the original projectile ion, having lost its charge, retains most of the translational energy. The effect is that the positive charge is continually being brought to rest, moving only a short distance between collisions. This explains the attenuation of radio waves by ions. If the ion moves between two electrodes with a potential difference across them it will make a large number of collisions before it reaches the negative electrode. If during this journey the polarity of the electrodes is reversed, the ion will begin to retrace its journey. If the polarity is continually reversed, and with increasing frequency, there will be a certain frequency of reversal at which the ion does not move far enough to encounter a molecule before its motion is reversed. At this frequency no energy will be lost by the oscillating ion to the surrounding gases. A radio wave may be considered as an oscillating electric potential gradient, and so at this frequency there will be no energy loss or attenuation of the radio signal. For small ions this frequency is about 30 MHz, but because electrons are so much lighter the fading due to electrons in gases does not disappear until the frequency of oscillation reaches that of infrared light.

While the concentration of charged species can be investigated using electric probes or radio-frequency absorptiometers, the only general method of identifying ions in flames is mass spectrometry.

In order to adapt a mass spectrometer for the study of ions in flames, a new type of inlet probe has to be used. A sample of flame gases is withdrawn through a small orifice into a chamber, where the pressure is considerably reduced by pumping the interspace with a high speed rotary pump. In this chamber the ions are accelerated towards a metal diaphragm in a strong potential gradient. They pass through a hole in the centre of the diaphragm and into the mass analysis system of the instrument, which can be of the magnetic sector or quadrupole type. At the same time most of the uncharged gas molecules are pumped away at right angles to the line of flight of the ions. In practice, the inlet system is more complicated than this and requires more than one diaphragm and pumping system. It may also be necessary to provide retarding plates with a positive potential imposed upon them to reduce the initial high velocity of the ions to a value which can be accommodated by the mass analysis system. However, no ion source is required and no transverse beam of electrons. It is merely necessary to collimate the ion beam with a series of apertures, and to reduce the pressure to a value sufficiently low to eliminate scattering of the collimated beam. Calibration of the mass scale is carried out empirically by adding substances to the flame which can produce ions of predictable and unambiguous mass. The alkali metal salts are ideally suited for this purpose. These can give a calibration in the mass range of 6 to 133. This is sufficiently high for the determination of the masses of most ions found in common flames.

4.3.7. *The pneumatic probe*

The probe which connects the ion source of the mass spectrometer to the flame gases may be modified so that it can provide, in addition to the measurement of the concentration of separate flame gas components, a determination of the local flame temperature. The principle is based on the concept of sonic flow in an orifice. If the gas pressures on the two sides of an orifice differ by a factor greater than two, then the gas travels with the speed of sound from the high pressure to the low pressure region through the orifice. The pressures are then related to the temperatures on either side of the orifice, the average relative molecular mass of the gas, and the Reynolds number of the gas flow. (The Reynolds number is the product of the average gas velocity and tube diameter divided by the kinematic viscosity of the gas.) If two such orifices are connected in series and the pressure differentials across both are separately greater than the factor of two, the relationships between pressure and temperature may be absorbed in a single empirical constant K, and:

$$T_1/T_2 = K(p_1/p_2)^2,$$

where p_1 and p_2 respectively are the pressures at the entrance to the

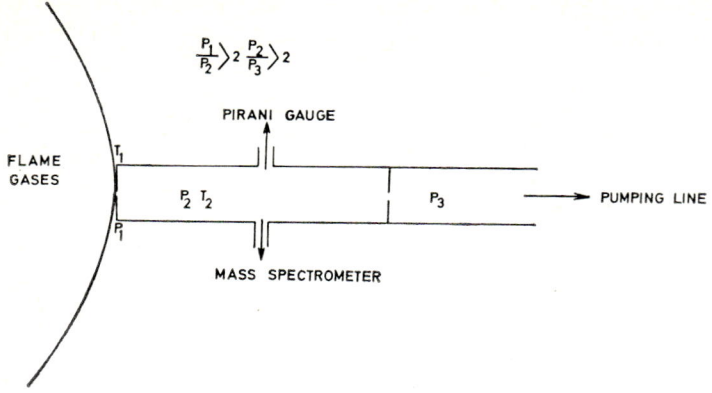

Fig. 4.5. Diagram of the pneumatic probe.

first orifice and between the first and second orifice and T_1 and T_2 are the equivalent temperatures (fig. 4.5).

In the practical realization of this concept the first orifice is the end of the probe which withdraws the sample. The space between the two orifices is connected to the mass spectrometer, and also to a second pressure-measuring device such as a Pirani gauge. The second aperture is connected to a conventional pumping line. The assembly must first be calibrated, using gas samples at known temperatures, in order to evaluate the constant K. The pressure p_2 can be read directly on the vacuum gauge, and is also available by summing the partial pressures of the components measured by the mass spectrometer. The relationship is only valid and K only remains constant if there is no large concentration of free radicals in the sample of gases withdrawn from the flame. As radicals disappear most readily by recombination on the walls of vessels, a high concentration of radicals, which would undergo recombination in the first orifice, would result in a change in the average molecular mass of the gases on passing through the orifice, thus invalidating the relationship. In order to minimize a time lag due to the pumping of gases through small apertures, the chamber between the two orifices must be made as small as possible.

Within the limitations imposed by these requirements, the technique provides a unique simultaneous measurement of concentration and temperature, and the results obtained are in reasonable agreement with those given by more conventional techniques such as the use of thermocouples or emission spectrophotometry.

CHAPTER 5
further applications

IN this chapter the role which the mass spectrometer has played in three quite different fields is examined; the fields of geology, medicine and space studies.

The use of mass spectrometers in geology has a comparatively long history, and their application to the study of samples brought back from the Moon was merely an extension of well tried methods used with samples of terrestrial origin.

The application of the mass spectrometer to medicine is, however, much more recent, and provides one of the growing points of the subject. The ability of the instrument to identify complex organic substances of biological origin, when only the very tiniest samples are available, is attractive to the clinician. Increasingly, he wishes to know more of the complex chemistry of his patient's body. Further, the ability of the machine to detect the presence of stable isotopes makes it possible to study metabolic processes by giving the patient substances labelled with perfectly safe stable isotopes such as ^{15}N or ^{13}C. The mass spectrometric technique is however still not sufficiently simple and reliable that it may be used routinely in the pathology laboratories of hospitals for the analysis of large numbers of samples. Whether we may eventually find the mass spectrometer playing this role is still doubtful. Some progress along this road has been made by the use of solid state electronics for modern instruments. There still remains the great expense involved in the purchase, maintenance and operation of these machines.

The use of mass spectrometers for space studies has been restricted by the sheer size of the equipment. Samples from the Moon have been examined back on Earth, but much more information could be obtained by actually placing mass spectrometers in space or on the surfaces of other bodies, and the first steps in this direction are only now being taken with the Viking missions to Mars.

5.1. *The mass spectrometer in geology*

The application of the mass spectrometer to geological problems dates from about 1935 when the great pioneering mass spectrometrist, Nier, demonstrated that all the radioactivity due to potassium was concentrated in the ^{40}K isotope, a discovery which was used later in the dating of the Earth's crust. The potentialities of the mass spectrometer in geochronology, or dating, were largely recognized by Nier and his co-workers in the period immediately before the last war. Today it is the standard instrument for determining the age and composition of minerals. It is assumed that the chemical nature of the Earth is due to the survival of the elements having the most stable nuclei; stable, that is, in a cosmological sense. This is borne out by the fact that most of the mass of the Earth is made up of a few elements such as iron, oxygen, silicon and magnesium. There is a considerable difference between this state of affairs and that in extraterrestrial matter where, for example, helium can be a major constituent. Helium was in fact first observed in the Sun, and only later identified upon Earth, where it exists as a trace element.

Of more direct interest to geology is the isotopic composition of the elements. When it became possible to measure the abundance of isotopes with sufficient accuracy, it was found that to a first approximation this composition was remarkably constant whatever the sample source. This led scientists to speculate that the isotopic abundance was related to the stability of the nuclei. Thus in polyisotopic elements it was found that the most abundant isotopes had even numbers of protons and neutrons in their nuclei. Whatever the origin of the sample, whether from Earth or from other parts of the solar system, the isotopic constitution is very nearly the same, suggesting that all this matter had the same history, i.e. that at one time it all had experienced the same effective high temperature.

Where there are variations in the isotopic pattern of the elements, this can be traced to the occurrence of some nuclear reaction, either as a result of natural radioactive decay or induced by bombardment with high energy particles from space. It is the small change in isotopic composition which is used to determine the age of geological specimens. It is only those naturally radioactive elements with a very slow rate of decay which can be used for making measurements on such a time-scale. Over the period required for their collection, treatment, and study their composition must remain the same, but it must change significantly in geological time. It is the intervention of nuclear disintegrations which is considered to be the main cause of the small change in isotopic composition with origin.

Minerals often contain small amounts of gases occluded in small voids within their structure, and it is not difficult to isolate the gases and to examine them with a mass spectrometer. These gases frequently contain substantial amounts of the noble gases, which survive readily because of their great chemical stability or inert nature. Helium is produced in any nuclear reaction which involves the emission of an alpha particle. Argon is produced by the decay of ^{40}K, krypton and xenon arise by fission, and radon is the result of uranium disintegration. Helium contains two naturally occurring isotopes ^{3}He and ^{4}He. The first is believed to arise by the decay of tritium, which is formed in turn by neutron bombardment. Neutrons in cosmic rays convert nitrogen in the upper atmosphere into carbon and helium. Underground, the ^{3}He is produced by neutrons released during uranium fission reacting with lithium. In one of his early studies Nier found that lithium minerals, such as spodumene, contained an excess of the light helium isotope. Argon-40 is the product of radioactive decay of potassium; not only is the isotopic constitution different in the atmosphere and in minerals, but the amount found may give an estimate of the age of the potassium-bearing mineral. The occurrence of krypton and xenon in rocks gives evidence of naturally occurring fission reactions. These may be induced by neutron bombardment or may arise spontaneously. The rate at which uranium and thorium decay by spontaneous fission is, however, very slow even on a time-scale which is long compared to the age of the Earth.

In addition to the variation induced by nuclear reactions, there are some changes in isotopic constitution which are caused by some natural fractionation process which is mass-dependent. Thus if the gases had to diffuse through a porous medium, such as sand or soil, the lightest isotope would diffuse fastest and might concentrate preferentially in a particular environment. More interesting, however, is the effect of the isotopic mass on the rates of some chemical reactions or the position of some chemical equilibria. Such equilibria are of course temperature-dependent, and a case of particular interest involves the isotopes of oxygen.

Carbon dioxide in the atmosphere is in equilibrium with the corresponding carbonate ions in sea water, but the equilibrium constants for the two isotopic species have different temperature coefficients. Thus the isotopic constitution of carbonate ion in sea water is dependent on the water temperature, and measurement of its composition gives an estimate of the water temperature within 2°C. The isotopic composition of the carbonate ion in the water is reflected in the composition of calcium carbonate laid down in their skeletons by marine animals. In a classic experiment, the ^{18}O/^{16}O ratio was determined in the fossil belemnite from the Jurassic period. It was found that this ratio changed in a periodic manner in successive layers. Thus not only was the temperature of the sea water 300 million years ago measured as accurately as with a common laboratory thermometer, but also the seasonal variation of the sea temperature was clearly demonstrated.

A further example of the fractionation of isotopes is given by the evaporation of water from the surface of seas and lakes. Because of the higher mass of heavy water molecules, D_2O and $H_2{}^{18}O$, these tend to concentrate in the liquid phase, so that the water vapour is richer in $H_2{}^{16}O$. The situation is restored if this falls as rain and rejoins the original water masses. If however it falls as snow on the Greenland ice cap or other such polar regions, it is trapped. As a result, ice from these regions tends to be enriched in the lighter isotopes. Again, measurement of the isotopic composition of ice field layers can give some information about past climates.

Another example of fractionation due to an isotope effect is provided by the variation in the relative abundances of ^{12}C and ^{13}C. The latter is perceptibly more abundant in sedimentary rocks, such as limestone, than it is in the naturally occurring carbon of animals and plants. This should not be confused with accumulation from the atmosphere in living matter of radioactive ^{14}C, a phenomenon which is used for the dating of archaeological objects made from once-living material containing carbon. The carbon dioxide which is incorporated into plants as a result of the photosynthetic process is enriched in the lighter carbon isotope, because of the higher rate constant for the lighter carbon dioxide. The concentration of ^{13}C in the skeletons of marine animals and plants which ultimately forms the chalk or limestone is also due to an isotope effect in the exchange of carbon dioxide and carbonate ion in the sea water. There is a further difference between marine and land plants: in the former the carbon dioxide is supplied in solution, while in the latter it diffuses as a gas from the atmosphere into the leaves of the plant. This is reflected in the $^{12}C/^{13}C$ ratios.

The case of the isotopes of sulphur is also an interesting one. The $^{34}S/^{32}S$ ratio is remarkably constant in samples derived from the meteorites which bombard the Earth, and it is supposed that this was the ratio of the isotopes which was predominant when the solar system was formed. Any variation in this ratio in samples of sulphur occurring on Earth is attributed to the effect of life. Bacteria are capable of reducing the sulphate ion, and this process goes on in many lakes in Africa. In fact during the Korean war when American native sulphur was in short supply, the possibility of exploiting this bacterial reduction was investigated. This reduction is associated with an isotope effect, the $^{34}SO_4$ being reduced at a slower rate. This means that the escape of hydrogen sulphide enriched in the lighter isotope concentrates ^{34}S in the residual sulphate in solution, and this may ultimately appear in sedimentary rocks. At the same time the precipitated elemental sulphur is enriched in the lighter isotope. Thus by studying the isotopic constitution of sulphur-bearing rocks, some idea of the biological activity going on during their formation may be assessed.

5.1.1. *Dating*

The principle of isotopic dating of a rock or mineral is based upon there being within that rock an element that changes in an identifiable manner with geological time. The time-scale on which it is required to operate may be gauged from the following dates. The age of the solar system is about 4.5×10^9 yr; the oldest rocks known are found in Greenland and are 3.7×10^9 years old, while the oldest rocks in Great Britain are found in N.W. Scotland and are about 2.7×10^9 years old. Basalt rocks brought back from the Moon have an age of about 3.3×10^9 yr, showing that at the time when these basalts were being laid down on the Mare basins of the Moon there were already on Earth rocks which had crystallized and retained their shape. A further requirement for successful dating is that the parent radioactive element must be detectable in the rock, while the daughter isotope produced by the decay process must be readily distinguishable not only from its parent but also from other naturally occurring elements within the rock. As an example, the decay of ^{40}K to ^{40}Ca has a suitable half-life, and the ^{40}K may readily be separated from the ^{40}Ca. However, the rate of production of this daughter isotope is completely masked by the huge excess of naturally occurring calcium. In addition to knowing the concentration of parent and daughter in the rock, the geologist must also know the rate constant for the radioactive decay, and for this he is dependent on the physicist.

The decay of ^{87}Rb to ^{87}Sr by beta emission has a half-life of about 5×10^{10} yr. Most igneous rocks contain traces of rubidium and if in an idealized case they originally contained no strontium, then their age could be determined directly from the amount of the single strontium isotope which had accumulated. However, most rocks contain some strontium as well as rubidium, and so it is necessary to measure the distortion of the normal isotope ratio by the accumulation of ^{87}Sr. It is usual to use the ^{86}Sr isotope as the standard and to express the ^{87}Sr concentration as a percentage of this. By examining numbers of rocks of different ages, a good estimate of the original ratio in the rocks may be made. Ages of about 10^9 yr have been measured in this way. The main disadvantage of this method of dating is that the amount of strontium found in most rocks is extremely small (183 p.p.m. in common granite), and this has led geologists to prefer dating based on the decay of uranium to lead, which is more plentiful and widely distributed.

One of the most useful dating schemes is based on the decay of ^{40}K to ^{40}Ar, but here we have a further complication. It was mentioned above that the decay of ^{40}K to ^{40}Ca by beta emission could not readily be observed because of the large excesses of ^{40}Ca found in most rocks. However ^{40}K also decays to ^{40}Ar by a process known as K capture (the K here referring to the innermost electron shell and not to potassium). The rate of decay by the latter process is only about a tenth of the former, but before an age determination may be made it is necessary to know accurately the exact ratio of the rates of these two decay processes as well as the half-life for ^{40}K. In ^{40}K dating the critical measurement is that of the concentration of ^{40}Ar within the rock. There is always the suspicion that some of the argon formed in the decay process has escaped by diffusion, while atmospheric argon may have diffused in, either naturally or during the sampling procedure. It is therefore necessary to measure precisely the abundances of all the argon isotopes in the sample of gas obtained from the rock. Only the ^{40}Ar isotope is produced in the decay process, but atmospheric argon contains ^{36}Ar and ^{38}Ar. The amount of these isotopic species in the sample is a measure of the invasion of atmospheric argon. The success of such a dating scheme is very dependent on the rock remaining undisturbed; when there is an intrusion of hotter, younger rock (possibly in the molten state) into a cooler, older rock, considerable disturbance occurs. This is reflected in the manner in which the geological age apparently changes in a continuous fashion across the boundary between the old and new rocks.

The exploration of the Moon's surface and the bringing back of a considerable number of rock samples gave a great stimulus to the development of methods of measuring isotope abundances by mass spectrometry. The standard technique is to use a thermal ionization source in which the sample is painted directly on the filament and leaves it in the form of positive ions. Although this method is extremely economical in sample consumption, at least 10 μg is required, and this is too large for many samples. Attempts were therefore made to increase the efficiency of ionization, and the most successful of these involved the use of activators, such as borax or silica gel, which if painted on the filament together with the sample increased the rate of ionization one hundredfold. As an example of the results achieved in recent studies, a sample was analysed which contained a total of 0.56 μg of rubidium and 0.9 μg of strontium. The ^{87}Sr/^{86}Sr ratio in this sample was determined as 0.71009 ± 0.00005.

It is, however, the uranium–lead system which has been used most widely in dating rocks on Earth, and with the use of activators the sample size can be reduced to as little as 100 ng. There are four stable isotopes of lead, and the average isotopic composition of naturally occurring lead is ^{204}Pb 1·48%, ^{206}Pb 23·6%, ^{207}Pb 22·6% and ^{208}Pb 52·3%. Of these, the three heavier isotopes are produced as the result of the radioactive decay of uranium and thorium, but ^{204}Pb is not so formed and may be considered as primeval lead, that is in the same state as it was when the Earth's crust formed. The processes producing the three radiogenic isotopes are:

$$^{238}U \rightarrow {}^{206}Pb + 8\,{}^{4}He^{++} \quad t_{1/2} = 4\cdot 5 \times 10^9 \text{ yr}$$
$$^{235}U \rightarrow {}^{207}Pb + 7\,{}^{4}He^{++} \quad t_{1/2} = 7\cdot 07 \times 10^8 \text{ yr}$$
$$^{232}Th \rightarrow {}^{208}Pb + 6\,{}^{4}He^{++} \quad t_{1/2} = 1\cdot 39 \times 10^{10} \text{ yr}$$

Thus the identification of ^{204}Pb in a sample shows that the rock contains some primeval lead, while its absence shows that all the lead has been formed by radioactive decay. As would be anticipated, most samples contain all the isotopes, and it was Nier who first showed that in common lead samples there was a considerable variation in the isotopic composition. This variation permits the geologists to estimate the age of the lead sample. It is assumed that at the time when the Earth was fluid, the composition of the Earth's mantle was uniform, so that the isotopic ratio for lead was invariant and this is called the primeval isotope ratio. When the crust solidified there was formed a local concentration ratio in the rock for uranium, thorium and lead. From then on the only change in this ratio came about by the radioactive decay of uranium and thorium, and was reflected in the change in the isotopic composition of the lead. It is not assumed that the radioactivity started only when the crust solidified. In fact it was going on since the original formation of the elements, but in a fluid state the results of decay were averaged out over the whole mantle and so could not be observed. A value for this average composition, that is the isotopic ratio for primeval lead, may be found from an examination of lead-bearing minerals which do not contain uranium or thorium. With this information, and with a knowledge of the chemical constitution of a sample containing uranium, thorium and lead, it is a simple matter to compute the age of the sample. Further, if the age can be checked by an alternative method it is then possible to determine the age of the Earth's solid crust.

The identification of primeval lead is, however, not simple; and although in the past it was identified as that lead which has the highest ^{204}Pb content, the uncertainty has persuaded geologists to make more indirect calculations. The need for such a method was strengthened by the observation that the lead found in meteorites had a higher ^{204}Pb content than any found on Earth, and also by the suspicion that many of the chemical analyses of minerals gave incorrect ages because of the preferential leaching out of one of the three elements. This alternative method is based on the fact that ^{238}U and ^{235}U have measurably different half-lives, and so ^{206}Pb and ^{207}Pb are being formed at different rates. Thus the ^{207}Pb/^{206}Pb ratio changes with geological time, and a measurement of this ratio and that of ^{235}U/^{238}U can be used to determine geological age. Ages of many uranium-bearing minerals have been measured to be of the order of 5×10^8 yr. Such a measurement is not invalidated by leaching, and in any case is more accurate than the corresponding measurement of chemical composition. The concentration of ^{204}Pb can still be used to correct the original chemical constitution. It is interesting that the knowledge of the isotopic composition of lead from different mineral sources, obtained for dating purposes, has been used recently for an entirely different end. The isotope pattern is characteristic of the mine of origin, and lead appearing as an environmental pollutant has been traced to its origin by its isotopic constitution.

5.1.2. *Ages of meteorites*

There is an increase in interest in the study of the isotopic constitution of the elements in meteorites, and while at present they represent the only samples available from the outer regions of the solar system, there is little doubt that space programmes will supplement them in the not-too-distant future. The age of meteorites as determined by the uranium/lead or rubidium/strontium method is of the order of 4.5×10^9 years, and this represents the time since solidification. Age determination by the potassium/argon method gives a shorter period (5×10^8 yr), but we are not necessarily measuring the same thing. While the first two methods, which are remarkably concordant, measure the age since solidification, the third may well measure the age since the meteorites cooled sufficiently to prevent the escape of radiogenic argon.

Before arriving on Earth, meteorites have spent a long time presumably in orbits within the solar system, and during this time they have been continuously bombarded by cosmic rays. As a result nuclear transformations are induced in the outer layers of the meteorites by the impact of the high energy protons of the cosmic rays. The resulting distortion of the isotope ratio gives some information about the intensity and period of irradiation. As occurs on Earth, lithium within the meteorites is converted into ^{3}He, and this is reflected in the change in the lithium isotopic composition and also that of the helium. Similar variations are revealed in the potassium isotope ratio, and the isotopic composition changes with distance of the sample from the outside surface. This shows the attenuation of the cosmic rays as they penetrate into the interior of the meteorite. While variations of this composition can be demonstrated by melting the sample in a vacuum and analysing the gas, or by painting the extracted sample on a thermal ionization filament, an interesting new technique is now being developed which obviates such lengthy sample handling.

In this new technique, which is particularly suitable for the study of surfaces, a beam of ions is directed on to the surface of the specimen. It is important to use a bombarding ion which is not of analytical significance. The ion beam penetrates only the outermost layers of the specimen, and the kinetic energy of the projectile ion is transferred to atoms in the lattice. As a result these are ejected, and may be collected in a mass analyser system and identified. While this sputtering process, as it is called, is very complex and leads to the formation of free electrons, positive and negative ions and neutral atoms with wide ranges of kinetic energy, this is not important for the analytical determination. It is sufficient to know that by this technique ions are produced which are representative of the outermost layers of the specimens (1 nm thick). Successive layers may be removed by continued bombardment, and thus reveal changes in composition with depth.

5.2. *The mass spectrometer in space*

Until the last decade there were only two methods of gaining knowledge about the composition of extraterrestrial matter: examination of objects trapped in the Earth's gravitational field and surviving passage through the atmosphere, and analysis of the light emitted by luminous bodies. While the examination of meteorites has not on the whole been very revealing, the spectroscopic study of the light coming to us from outer space has been very fruitful, with such successes as the discovery of helium in the Sun's atmosphere, the detection of carbon dioxide around the planet Venus, and the demonstration of the presence of hydrocarbons in the heads of comets.

Due to the great advances in rocketry, however, it is now possible to consider not only sending men to the Moon to bring back samples of extraterrestrial material for study upon Earth, but also to despatch instruments to all parts of the solar system to send back information about its composition and structure. One of the instruments eminently suitable for consideration in such a programme is the mass spectrometer, and a detailed list of the future extraterrestrial applications of the instrument has already been compiled. In the present stages of lunar exploration the mass spectrometer has been recommended for a number of investigations.

The highly rarefied lunar atmosphere could be analysed and monitored continuously by a small instrument operating on the surface of the Moon, or from a small orbiting satellite. A further development would be a small hand-carried unit which could be used to detect local concentrations of gases caused by, for example, volcanic activity. A less likely possibility is the screening of lunar materials before their transport to Earth. At the time of writing every specimen brought back by the Apollo lunar exploration programme is of the greatest possible scientific interest, but future astronauts may be required to be more selective. In the February 1971 Apollo operation, attempts were made to sample the solar wind on the Moon's surface. All the inner planets are within the Sun's ion atmosphere and are bombarded continuously with charged particles. This flux could be sampled and monitored by a remote instrument, either carried in a space vehicle or left upon the surface of the Moon.

During the nineteen seventies unmanned spacecraft carrying instruments have been travelling to many parts of the solar system sending back new information about its origin and composition. Unmanned space vehicles have already visited Mars and Venus. The Viking probes are at the time of writing on the surface of Mars and the first mass spectra of the Martian atmosphere have been received. They show that all the major components of the terrestrial atmosphere are also present on Mars, but the relative proportions are quite different. The most abundant gases are carbon dioxide and nitrogen, and it is interesting that the isotopic composition of the Martian nitrogen, that is the $^{15}N/^{14}N$ ratio, is different to that of terrestrial nitrogen with a greater concentration of the heavier isotope. This suggests that at one time the atmosphere on Mars was far denser than it is today. One of the most important tasks of such probes is the search for evidence of life outside Earth, either as living tissue or fossil remains. For this work more sophisticated machines will be required, capable of detecting volatile organic molecules. Of greater interest to the cosmologist, however, will be the studies of the occurrence of atoms, small molecules and ions in interplanetary space and in the heads of comets.

Further valuable evidence on the age and origin of the planets would arise from a study of the isotopic composition of surface rocks. The rates of decay of the long-lived radioactive isotopes are known, and estimates of the age of rocks may be made from measurements of the local concentrations of such isotopes and their daughter decay product atoms. Finally, there is the possibility of determining the constituents of planetary atmospheres and their variations. In some cases, such as the satellites of the major planets, these atmospheres may be very thin and no inlet pumping system will be required; but on others, such as the planet Venus, the atmospheric pressure will be too high to allow direct admittance to the ion source.

For each of these applications it may be necessary to design a different mass spectrometer, but experts predict that only two general types will be required. For the measurement of the constituents of rarefied atmospheres a comparatively simple instrument is sufficient. It need only have a mass resolution of between fifty and one hundred, and need have no inlet system; there will be direct access to the ion source. A high sensitivity is required as the pressures fall to as low as 10^{-10} N m^{-2}, and the abundance sensitivity should be at least one thousand. The time required to scan a mass spectrum need not be shorter than, say, five minutes unless the instrument is to traverse the atmosphere at high velocity. The same mass analysis system could be used in conjunction with a small heating device to evaporate the volatile constituents of rock samples, either organic or inorganic. Once again no pumping is required. In a further application the ion source itself could be replaced by focusing electrodes, to permit the study of ion atmospheres.

The second type of instrument would have to be more elaborate, as it should be able to identify moderately complex organic compounds. A mass resolution in the range of five hundred to one thousand is necessary. In addition, some form of sample introduction system would be required, with the facility of vaporizing relatively involatile samples. For ionization an electron bombardment source would be most generally useful, but if in addition the same instrument were also required to analyse rock samples and to make isotope determinations, then a spark source would be required and a very much higher abundance sensitivity of about one million.

For the low resolution instrument without ion source the quadrupole design would seem to be most appropriate, because of its low weight, simplicity and cheapness. For the more complex higher resolution machine requiring an ion source, the most attractive design would appear to be the double focusing geometry of Mattauch and Herzog in either of its configurations. Miniaturized models with small permanent magnets have already been constructed and tested.

5.3. *The mass spectrometer in medicine*

It is probably only in the past ten years that mass spectrometry has been recognized as being of value in clinical medicine. Strangely enough it was not the high resolution or extreme sensitivity of the technique which encouraged its initial adoption, but the ability to carry out one of the simplest tasks, the quantitative analysis of air. But it was the air inside the body that the clinician was interested in, for by a study of the composition of the air in patients' lungs the efficiency of respiration could be assessed.

The significant gases are of course oxygen and carbon dioxide, but it can also be useful to measure nitrogen and argon and certain anaesthetic gases. It is not sufficient merely to analyse a sample of expired air, for respiration is a dynamic process and it is its rate which is important. This requires the mass spectrometer to give a continuous measurement of the lung gases. This has been achieved by passing a heated capillary tube into the patient's trachea or alveoli. The gases pumped along this capillary tube were led into the ion source of a low resolution mass spectrometer. The first instruments used were 180° focusing machines with a permanent magnet, and the ion accelerating potential was altered repetitively in steps to bring in turn to the collector the currents due to the ions of oxygen and carbon dioxide. In commercial models an oscilloscope gives a continuous visualization of the state of the respiratory process. Other designs have used sector instruments with separate collectors placed at the focus points for the significant ion currents. Such instruments have also been modified to measure not only lung gases but also blood gases. This can give information about cardiac output and circulation. The blood gases are allowed to diffuse through a fine membrane placed at the end of a probe which is inserted into the blood vessel. After diffusion through the membrane they are once again conveyed to the ion source through a capillary tube. In addition to its clinical use, the respiratory mass spectrometer has also been used to study the breathing of normal subjects under adverse conditions, as in unpressurized aircraft cabins or in space capsules.

Another example of the application of mass spectrometry to medicine is in the detection and measurement of trace levels of metals in tissue and body fluids. The interest arises in two ways; first, because of the ever-rising levels of poisonous metals such as lead and cadmium in our environment; and second, because of our increasing knowledge of the important part which trace metals play in enzyme systems. For both these purposes it is necessary to measure amounts of metals at levels in the parts per billion range, and the spark source mass spectrograph is the only instrument capable of making such measurements on all metals at the same time. The technique has been used to determine the trace elements in teeth, tumour tissue, and in hair, and more specifically to determine thallium concentrations in body organs. While some metals can also be assayed at below parts per million with other chemical techniques, such techniques lack the ability to distinguish between harmless and lethal isotopes, such as those of uranium.

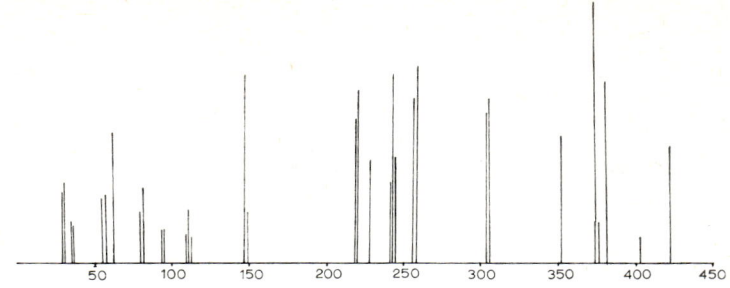

Fig. 5.1. Mass spectrum of mixed organic components of human blood.

One of the most attractive future prospects for mass spectrometry in medicine is the assay of drugs in tissue and in biological fluids. Treatment of patients who have taken accidental or deliberate overdoses of unknown drugs requires the ability to identify the drug and its approximate level. Fig. 5.1. shows the mass spectrum obtained from a sample of normal blood. Sometimes the spectrum of an ingested drug may be so specific that it can be disentangled from the spectrum of the mixture. It is more common, however, for a prior separation procedure based on ion exchange or thin-layer chromatography to be required.

Another important use of mass spectrometry in medicine is the metabolism of drugs, a discipline known as pharmacology. Here the rate of appearance and disappearance of drugs in various parts of the body is studied, and the metabolites which are formed are identified. One field of great importance is the study of the pain-relieving drugs used in childbirth. It has been possible in some cases to trace the drug from its injection into the mother across the membrane and into the urine of the infant. There is now available a substantial library of the mass spectra of commercially available drugs and their metabolites, and with the aid of computer processing of data given by the gas chromatograph–mass spectrometer combination on stomach washings, the identification of drugs by spectrum matching is routine in some advanced hospitals.

One of the most important groups of biologically significant compounds is the steroids, and it is difficult to disentangle that work carried out on these compounds which has only biochemical relevance from that which is more clinical in character. Although steroids occur in all living matter both in the animal and plant kingdom, the majority of the mass spectrometric studies have been confined to steroids of mammalian and human origin. Here it is known that they play a profoundly important role in reproduction, in digestion, and in cardio-

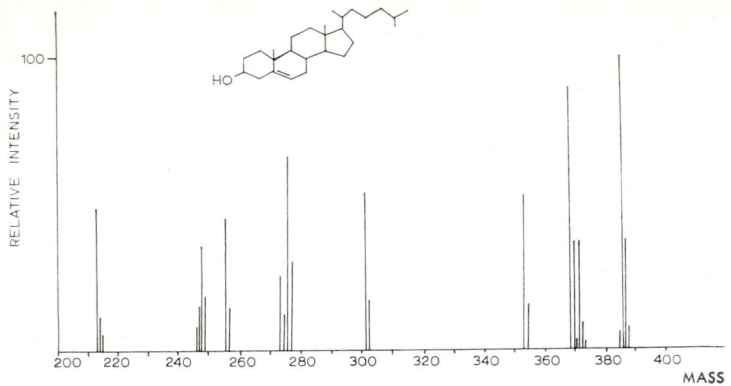

Fig. 5.2. Mass spectrum of cholesterol (relative molecular mass 386).

vascular disease. A typical steroid is cholesterol, a major constituent of animal fat, butter and milk. This is a compound of fairly low relative molecular mass, of moderate volatility, and with an easily identifiable mass spectrum. All steroids are variants on this basic four-ring structure (fig. 5.2), with differing side chains, numbers of double bonds, and substituent groups.

The mass spectrometer has played a dual role in clinical studies of steroids. It has provided a way of identifying individual steroids isolated by gas chromatography, and it has been used to determine the structures of unknown steroids or metabolites of known steroids. Because it would be difficult to deduce very much from a mass spectrum of a biological fluid containing all the normally occurring steroids together with, say, a steroidal drug, it is most common to use the gas chromatograph–mass spectrometer combination to separate each steroid before admission to the ion source. As the volatility of many steroids is not sufficient to allow them to pass through the gas chromatograph at the comparatively modest temperatures which are permissible, it is common practice to increase volatility by converting polar groups into non-polar groups before injection into the gas chromatograph. Thus OH groups can be converted into trimethylsilyl ether groups:

$$-\text{OH} \rightarrow -\text{OSi}(\text{CH}_3)_3$$

Before any investigation into disease or dysfunction can be carried out, it is necessary to study the normal levels of steroids in biological fluids. Thus, if the normal outputs of the sex hormones in urine and also the change in these levels with ovulation and pregnancy are determined, it becomes possible to detect any abnormal variations occurring as a result of disease or the injection of drugs.

Steroidal drugs are now widely used as oral contraceptives, and many of the resulting metabolites have been identified from their mass spectra. The great advantage is the very high sensitivity. It is no longer necessary to obtain visible amounts of the compound; most identifications can be carried out on the microgram scale. Chemical reactions can be carried out and the result of the reaction found from the mass spectrum of the products.

It is useful to know what the normal amount of a particular compound is in the animal or the human body. This is commonly called the body pool. In small mammals, for example, the water pool could be measured either by drying the carcass to constant weight, or more subtly by injecting a small amount of heavy water into the live animal, withdrawing a known volume of a body fluid, and measuring the isotopic composition of the water in that fluid. This is the familiar isotope dilution method and it can be applied to a variety of compounds. The pool of urea within the human body might be measured by the consumption of a small amount of urea labelled with ^{15}N. The isotopic constitution of the urea excreted then gives a measure of the urea body pool.

A few years ago an interesting medical debate was clarified with the aid of mass spectrometry. It had been shown that paper chromatography could be used to analyse the urine of patients suffering from mental disorders. When after chromatography the papers were sprayed with a reagent consisting of a mixture of 4-dimethylaminobenzaldehyde and ninhydrin, a characteristic pink spot appeared in some cases. This spot was particularly marked in the case of patients suffering from schizophrenia and was absent from normal controls. The value of such a simple diagnosis could not be questioned, but the identification of the chemical substance responsible for the spot proved more difficult. Finally, by comparison of the spot position with those of other compounds chromatographed under similar conditions, it was decided that the compound was β-3,4-dimethoxyphenylethylamine, a compound closely related to other amines known to exist in the human body. However, these catecholamines (dihydroxyphenylethylamines) would require an unlikely methylation process to take place before they could be excreted as β-3,4-dimethoxyphenylethylamine. Although attempts were made to correlate the mental state with the appearance of a pink spot from the urine, this postulate was not acceptable to a number of biochemists.

The problem was resolved by extraction of the spots from the papers and submitting the resulting extract to mass spectrometry. This revealed that the resolution on the paper had only been partial, and that the main substance responsible for the pink spot was p-tyramine (4-hydroxyphenylethylamine). The excretion of this compound was slight in normal patients, even those on a diet of foods containing large amounts of p-tyramine, such as cheese. The amounts of the compound in the urine of mentally disturbed patients were much higher, and this was most marked in the case of patients suffering from Parkinson's disease. When by the administration of drugs these patients experienced temporary relief, p-tyramine excretion fell to normal, but on a relapse the level rose again. The story was completed by the detection of β-3,4-dimethoxyphenylethylamine as a minor constituent of the pink spot, and also at a trace level in tea.

CHAPTER 6
current trends in mass spectrometry

6.1. *The mass spectrometer and the computer*

EVERY mass spectrum presents an enormous wealth of data: the relative molecular mass of the compound and in some cases its state of aggregation, its atomic constitution, and the isotopic composition of the elements from which it is formed. We are also given information about the many chemical reactions which take place when the highly excited molecule ion dissociates, and the atomic and isotopic composition of the end products—the fragment ions. Finally, there is information about the strength of the chemical bonds, and the kinetic energy released when they are ruptured. It is not surprising, therefore, that a great deal of this information goes largely unheeded because of the time taken for identification and interpretation.

The need for some automatic device for registering all this information and perhaps interpreting its significance has been recognized for some time, and this has led instrument designers into the difficult field of coupling the analytical power of the mass spectrometer with the data-processing capabilities of the small computer. Before such a programme was initiated, however, many mass spectrometrists had called to their aid the large computers to which many of them had access. One of the tasks which the computers were asked to perform was the prediction of the mass spectra of polyatomic polyisotopic molecules for comparison with recorded spectra. For example, it would be difficult to validate the mass spectrum of tin(IV) bromide without such aid. A second task was the determination of atomic constitution from precise mass. This can of course be carried out manually with the aid of a pocket electronic calculator, but even so the exact matching to six significant figures can be very time-consuming, and many programs have been devised to make the conversion automatically.

Such assistance, time-saving as it is, does not even begin to fulfil the requirements detailed above. What is required is a small computer dedicated to the output of the mass spectrometer alone. Then, in addition to recording the output data, the computer can be used in a reactive mode in which it usurps some of the functions of the operator and controls the instrument, so as to give maximum performance at all times. This need was demonstrated most graphically in a recent design survey for the computerization of instruments carried out in the Chemistry Department of Birmingham University. The intention had been to construct a time-sharing system in which a small computer would accept the data from all the instruments in the Department, storing them and processing them as time and storage capacity became available. It soon became obvious that the rate of data output from one high resolution mass spectrometer was greatly in excess of that provided by all other instruments, and it was suggested that no useful purpose would be achieved by such time-sharing.

Before the output from a mass spectrometer can be stored by a computer it must be converted into a form which is compatible with the data-handling systems of the computer, that is, it must be digitized. When the mass spectrum is being recorded in a conventional manner the ion currents, corresponding to the rates of formation of the various ions in the source, are collected after mass separation, amplified and perhaps allowed to flow through a galvanometer where they cause a deflection. This deflection rises to a maximum and then falls to zero (as the peak is scanned). The information provided by such a record is the steady ion current at a particular mass value; i.e. it is the peak height which is important. This is an analogue representation of the ion current. However, the computer cannot deal with d.c. signals like this, as its function is to count pulses.

The first stage therefore in the computerization is to convert these analogue signals into digital signals; peak height current or voltage is replaced by a number of counts, and this process is called digitization. The number of counts should be proportional to the peak height and therefore to the original ion current. It might be considered at this point that it would be sufficient to feed the ion current itself directly to the computer store, because after all an ion current is only a series of pulses corresponding to single ions arriving at the collector. However this simplification is made difficult because of the very short time-scale involved and the response time of most computers. Recourse is therefore made to digitization of the amplified ion currents, and when this is done the mass spectrum can be stored in the computer as a series of numbers at computer addresses (locations within the computer memory) identified as specific mass values. Such a form makes matching of spectra a comparatively straightforward operation for the computer. In order to carry out the digitization process continuously an essential part of the system is a peak sensor. The digitizer would otherwise continuously output counts from noise or baseline drift. Peak sensing can be carried out either on the original analogue form by differentiating the output and detecting any change in the sign of the voltage, or by a comparator circuit which continuously compares a sample voltage with that obtained previously and detects the peak maximum by a change in slope. In either case the signal is then fed to the digitizer. This may be one of a number of devices, the simplest being the voltage-to-frequency converter.

The next most important step is the identification of each of the mass peaks now stored as numbers of counts in different addresses of the computer store. To understand this it is necessary to return to the original manual examination of mass spectra and to remember that a mass spectrum is numbered by inspection. The operator uses his experience of any individualities of scan rate or recorder, and the blank spectrum of his instrument, to assign integral mass values to the peaks which he observes. That is to say, he compares unknown peaks with known peaks, and estimates their relative positions in the scan. This process has to be carried out by the computer. Some assistance in numbering scans has been obtained by the use of mass markers, that is devices which will mark the recorder trace at significant mass values. These devices function by measuring magnetic fields at which ions of known mass are focused on the collector, and comparing them with the magnetic fields at which the unknown ions are focused on the collector. Prior calibration is of course necessary, and one can use such a calibrated device to provide a signal for the identification of the unknown peaks in the spectrum.

However, it is more useful to avoid this calibration and carry out the comparison with the aid of the facilities provided by the computer. The calibrating substance is admitted to the source at the same time as that of the unknown compound whose spectrum is to be recorded. At the same time as the scan is initiated, a digital clock is started, and the time of arrival of each peak is now recorded together with its digitized peak height and stored. A programme is then initiated which searches the store for evidence of the calibration compound. Thus if the known compound was the commonly used heptacosafluoro-tributylamine $((C_4F_9)_3N)$, the programme would search for the large peak due to the ion CF_3^+ with an integral mass of 69. Having found this peak, the search is extended to other known reference peaks, say 81, and so the mass scale is calibrated, not in terms of recorder chart length, but of time. With such a calibrated scale the addresses of all the other counts in the store corresponding to the peaks in the mass spectrum can be identified in terms of their integral mass. This procedure can be carried out on line, so that even if the mass spectrometer is coupled to a gas chromatograph and information is being accumulated very rapidly, it is still possible to provide a print-out on a teletypewriter of the integral mass values and the corresponding peak intensities as numbers. It is only a further small refinement to extend the program to divide all the intensity numbers by the highest and provide a print-out which gives the cracking pattern of the compound. To aid the recognition of this pattern it may be drawn out in the form of a histogram or stick spectrum, by a plotter, or it can be visualized immediately on a cathode ray screen.

The operator may not desire to examine each spectrum, but merely to know whether the compound has been studied previously. The computer can then be programmed to carry out a library search in which the cracking pattern is compared with a library of mass spectra which have been stored either on magnetic tape or, more probably, on a computer disc store. Some thousands of spectra may be stored on a single large disc. Such a peak comparison, however, takes time and this time increases as the size of the library increases. The very large library held by the Mass Spectrometry Data Centre at Aldermaston will contain over a hundred thousand spectra by the time this book is being read. To cut down the time of the library search it is necessary to reduce the number of peaks compared. The program may either select the eight most intense peaks or, perhaps most rigorously, the two most intense mass peaks in each fourteen mass units (corresponding to the unit CH_2 in a homologous series).

If this search has been futile, and no match can be made with those existing in the library file, it is then necessary to gather more information about the compound in order to determine its structure. The first additional operation is to produce a high resolution mass spectrum, and to store the information, not in the form of integral mass values, but in the form of precise mass values. As the digital clock produces a pulse every microsecond, there is sufficient accuracy to define the precise mass in terms of its time of arrival during the scan. A further program can now be initiated in which the precise mass is compared with sums of multiples of the precise masses of the elements, and the atomic constitution is thereby identified. The print-out can now be in terms of ion abundance and atomic constitution.

The most recent refinement to the computer processing of mass spectral data is the application of artificial intelligence to the determination of structure. The ultimate objective of such an enterprise appears to be the complete automation of the interpretive process, so that the operator merely introduces the unknown sample and waits for his answer in the form of a print-out of the possible structure. We are still a long way from such a situation, and are likely to remain so for the foreseeable future, since the complexity of a computer program which could anticipate all the subtleties of the structure of polyfunctional organic molecules is likely to be too great to be envisaged. However, the computer can play a role in the interpretation of mass spectra, and its most useful function is its ability to test every possibility, using a number of different criteria without fatigue and without making mistakes. An example of what can be achieved is the identification of the class of compounds to which the unknown belongs, i.e. the compound type. This involves searching for hetero-atom content in the molecular formula and applying some empirical rules formulated in the prior study of compounds containing that particular hetero-atom. The application of such a program would be restricted to monofunctional organic compounds. It is this restricted role that may be expected of successful computer interpretation.

Let us take as an example the pair of compounds pentan-1-ol and 2-methoxy-2-methylpropane. Both have a relative molecular mass of 88, and contain a single oxygen atom. The presence of this oxygen atom within the molecule would be detected from a consideration of the precise mass, the carbon isotope ratios or merely the integral mass of the molecule ion. The rest of the molecule would then be identified as saturated and the search restricted to saturated oxygen compounds such as alcohols and ethers. The dissociation patterns of these two types of compound are quite different, the base peak for pentanol having a mass of 42 and that for the methyl ether having a mass of 45. The computer could thus compare observed spectra with patterns predicted from a prior study of the spectra of a large number of alcohols and ethers. Having assigned each compound to its class the computer could then provide more structural information, such as the length of the hydrocarbon chain.

It has been mentioned in the section dealing with the application of the mass spectrometer to organic chemistry that it is possible to determine the sequence of amino acid residues in a polypeptide. Only a few peaks in the spectrum are significant for this process, and there are only some 22 possibilities for naturally occurring amino acids. It is, therefore, comparatively simple to program the computer to search for peaks due to the successive losses of amino acid residues, and having made the assignment to check it by comparison of the measured and calculated precise masses of the molecule ion. Several groups of workers have successfully applied such programs to the elucidation of the sequence of amino acids in polypeptides with up to a total of fifteen units.

6.2. Possible technical advances

The mass spectrometer has grown in the past fifty years from a research experiment of basic physics to a powerful analytical tool which can provide useful data for all branches of science from civil engineering to botany. During this time its potentialities have been increased enormously, from a mass resolution of 10 to one of 100 000, and from a device for studying simple gases to one capable of dealing with complex organic and inorganic molecules. Three aspects of the mass spectrometer are theoretically capable of further improvements: the sensitivity, the resolution, and the mass range. These are of course not independent, and one may well achieve an advance in one facility at the expense of one or both of the others.

6.2.1. Sensitivity

The sensitivity of a mass spectrometer depends firstly on the minimum number of ions arriving at the collector which are necessary to define a mass peak in a given time. This number depends in turn on the resolution required; one must construct a peak with a sufficient number of ions to be able to define its centroid or centre of gravity. Using the present-day particle multipliers it is possible to record a mass spectrum at a resolution of about 1000 with a sample size of 1 ng.

Assuming that in order to define the molecule one has to record a mass spectrum with a maximum peak height ratio of 100, that is a factor of 100 between the ion current of the most abundant and that of the least abundant ion, then if the ionization efficiency and transmission were unity and a molecule with a mass of 600 were being examined, 10^{10} ions would arrive at the collector to define the smallest peak in the spectrum. However, our assumptions are incorrect by a huge factor, for the numbers of ions arriving at the collector are in the 10 to 10^2 range. In fact the ionization efficiency of most sources is only of the order of one part in 10^5, and the transmission factors are variable but low. What hope is there then of improving the situation?

The electron impact ionization source is already in an advanced state of development, and it seems doubtful whether any large improvements in sensitivity are possible. Already electron currents approaching a milliamp are being used, and moderate magnetic fields are used to lengthen the electron paths into spirals with consequent improvement in the chances of successful collisions with sample molecules. The further increase in the efficiency of ion production must await the development of an alternative ionization process, such as chemical ionization, with a sufficiently high cross-section.

The transmission of the ions through the instrument is restricted by the very system which provides the resolution, that is the slit and lens arrangements, and although improvements in sensitivity have been achieved recently in this field by adding hexapole electrode assemblies to maintain the rectangular cross-section of the ion beam, further improvements will be difficult and expensive. One way of economizing on sample would be to restrict the sample to the physical area bounded by the source slit, or by reducing the rate at which sample is pumped out of the ionization region during the recording of the spectrum. By a combination of small improvements on many of these factors it seems possible that the sensitivity could be raised by two or three orders of magnitude, but this would surely be the practical limit. In the meantime, however, some substantial increases in sensitivity can be achieved if the demands for information can be reduced; that is to say if we are satisfied with only that part of the mass spectrum which contains characteristic features.

Suppose that in the mass spectrum of a known compound one peak may be selected that by its precise mass value is sufficiently characteristic of the compound to be determined. The most suitable ion would in most cases be the molecule ion, but this may not be sufficiently abundant to make it a satisfactory choice. The instrument is tuned to record this ion current alone, and then the sample is admitted to the ion source. The signal would begin to rise as the sample began to appear, reach a maximum, and then decay as the sample was depleted, finally returning to zero. The area under the curve of ion current with time would then be proportional to the total number of ions produced during the time and therefore, one would hope, to the amount of sample passing through the source. If the sample is being evaporated from a direct insertion probe, the process is called the Integrated Ion Current method. If on the other hand the sample comes from the base of a gas chromatography column via a separator, then the technique is called Mass Fragmentography. The sensitivity of the method is highly dependent on the choice of the ion which carries a high proportion of the total ion current, but in the favourable cases of polycyclic hydrocarbons as little as 10^{-14} g can be measured.

The use of only one peak may lead to some uncertainty, and the situation is greatly improved if two or more ion currents can be recorded successively. This requires that the instrument should not scan continuously in the normal fashion, but should record separately the ion currents at several, say six, selected mass values. Such an instrument requires very rapid response and is best controlled by a small computer which also assembles the collected information. These

Fig. 6.1. Demonstration of the existence of isomers using the integrated ion current technique.

machines are in an active state of development, and the rapid response facilities of the quadrupole mass filters make them suitable for the task. Fig. 6.1. shows signal records for a single substance and for a pair of isomers. When using this technique for quantitative analysis the instrument must either be calibrated with a set of standards before the determination, or an isotope dilution technique must be used. As an example, nanogram amounts of p-tyramine (β-(4-hydroxyphenyl)-ethylamine) have been measured in brain tissue extracts by mixing the sample with a known amount of p-tyramine labelled with two deuterium atoms and comparing the integrated ion current curves at two mass numbers.

6.2.2. Resolution

The resolution of the mass spectrometer has increased by an order of magnitude for every decade since its invention. Can this continue or is there a theoretical limit? At the time of writing the resolution of commercial double focusing machines lies in the range 1–200 000, a recent demonstration showing the separation of the ^{12}CD, ^{13}CH doublet at mass 211 with a resolution of 136 000. Instruments operated by research physicists do appreciably better than this. The mass synchromer at Princeton is a special case, Smith having made some determinations with much greater accuracy, while Matsuda, working in Osaka in Japan, has achieved resolutions of one part in 400 000. Such measurements require that the ion accelerating voltage be stable to less than one part in 10^7. Since in general the position of a mass peak may be determined with a precision of ten times the instrumental resolution, masses are being measured to an accuracy of a few parts in 10^7.

Already it is necessary to take account of the mass of the electron, the difference between the mass of the molecule and the singly charged ion being of the order of 1 part in 10^6. However, further increases in resolution may have to take into account those tiny changes in mass which result from changes in the energy states of extranuclear electrons. Already the mass spectrometer has revealed structure within the nucleus. If one follows Aston's example and constructs a 'packing fraction' curve using all the new information about the large number of nuclides collected since his day and also using modern precise mass values, one notices sharp discontinuities, the most obvious being around the 80-neutron region. This is most clearly seen if one measures the mass difference between a pair of isotopes two neutrons apart, as for example ^{35}Cl and ^{37}Cl, and compares this difference with the mass of two neutrons. The mass discrepancy is called the neutron pair separation energy, and varies along the curve, showing the existence of closed shell structures within the nucleus. The energies binding protons and neutrons in the nucleus are large, but the same principles apply to the shells of electrons around the nucleus, and the expected difference in mass can be estimated by using Einstein's familiar equation $E=mc^2$. The difference in mass between the ground and excited states of an atom is of the order of 1 part in 10^{10}. Another energy change would be the formation of molecules from atoms, the pairing of the electrons in a chemical bond leading to a loss in energy and hence a reduction in mass. In the case of the hydrogen molecule the mass deficiency would be about 4×10^{-9} mass units requiring a mass resolution of one part in 5×10^8.

Although we are some way away from resolutions of this order at present, they nevertheless offer us an attractive prospect—the possibility of investigating molecular structure from measurements of precise mass. It is of course very uncertain whether this end will be achieved, for close behind it comes the ultimate limit to resolution. Account has already had to be taken of the relativistic gain in mass by the ion as it accelerates under the influence of the high voltage carrying it to the detector. Now any sample of gas molecules will have a Maxwell–Boltzmann distribution of velocities in the ion source, the so-called thermal velocities. Any ion will therefore have superimposed upon its normal velocity (that velocity it picks up in the potential gradient) a thermal velocity, which will be random in magnitude and direction and will have a spread in energy equivalent to a mass difference of a few parts in 10^{12}. This effect will smear out any measurement made with resolutions below this level, and it presents the ultimate barrier.

6.2.3. *Mass range*

The other horizon of mass spectrometry is the limit to the mass range. There are two distinct factors contributing to this: the provision of sufficiently powerful magnetic and electrostatic sectors, and the problem of evaporating substances with a high molecular mass. The solution to these problems may be like the solution to those of resolution and sensitivity discussed above, the provision of a sufficient amount of money to overcome them. At present most commercial instruments have a mass scan at full accelerating potential which covers about 1000 mass units. Higher mass ranges are selected by reducing the accelerating potential in steps with consequent loss in sensitivity and resolution. The indications are that higher resolutions require higher accelerating potentials, so that these two demands are opposed. The solution to the problem of the accurate mass measurement of ions with high mass values must therefore be the provision of larger and more powerful magnetic sectors with large radii and stronger magnetic fields, and this suggests the use of superconductive magnets.

The problem of producing the ions with high mass values is, if anything, more intractable. Today high molecular mass substances are transported into the vapour phase by simply raising them to the appropriate high temperature. The great advance with general purpose analytical machines was the invention of the direct insertion probe, a rod which could be pushed through a series of seals directly into the ionization chamber, with insulation safeguards to prevent the high source potential running to earth through the operator. Samples in a small tube at the end of the rod are therefore in position within a few millimetres of the electron beam, and upon heating, the evaporating molecules are rapidly ionized before meeting any surfaces to which they could adhere. With such a device, very involatile substances of high molecular mass can have their mass spectra recorded. However, the ease of evaporation or, to be more precise, the heat of sublimation, varies widely with the nature of the sample. While non-polar compounds like fluorocarbons can be evaporated at quite low temperatures —indeed some workers have claimed to have observed ions from tetrafluoroethene polymers with mass values up to 6000—quite low molecular mass polar compounds containing, for example, several hydroxyl groups as in sugars, cannot easily be evaporated. This has led to conversion of such polar compounds to non-polar derivatives, either by acetylation with perfluoroacid anhydrides or by making the trimethylsilyl ethers with trimethyl silylchloride.

It would however be far more convenient if substances of high molecular mass could be converted into gaseous ions directly. This is prevented in the case of, say, proteins or polysaccharides by the fact that the heat of sublimation is greater than bond dissociation energies within the molecule. That is to say, the attractive forces between the probe surface and the sample molecule are greater than the forces holding the atoms together in the molecule. Alternatives to heating may be dispersion from the probe surface with a high electric field, the use of high intensity laser light, or the introduction in solution into a complex molecular beam system. Both have been tried but further development is required. In the immediate future it would appear that the upper limit to the useful mass range will be about 2000.

This may not be quite so restricting as it appears, for the molecule ion peaks in the spectra of many large molecules are vanishingly small, the electron impact process being extremely destructive. The mass spectrometer could equally well be coupled to some other reaction cells in which the sample suffered reproducible photolysis or pyrolysis. The lower mass fragments resulting from breakdown could then be examined.

6.3. Conclusion

Whatever the outcome of developments in instrument design under the three headings quoted above, there will certainly be great advances in design which will increase our understanding of ionization and mass analysis. The new methods of observing the metastable peaks will continue to give more information about the dissociation of excited ions. The mechanisms and the energy released in specific dissociations have already been recorded for many cases. What is lacking at present is a knowledge of the structure of the ions leaving the ion source and striking the collector. Perhaps it is not too optimistic to envisage some methods of finding out directly how the atoms, which we detect by precise mass analysis, are arranged in the ion. Other techniques such as photoelectron spectroscopy and electron energy loss analysis are giving us more and more information about the ionization process itself. Perhaps in the future some combination of these techniques, which measure the energy in the scattered electron, will be combined with mass spectrometry, which gives us the information about the other product of the electron impact process, the ion.

APPENDIX
Some isotopic masses, abundances and ionization potentials

Atomic no.	Element	Symbol	Relative atomic mass	Mass no.	Precise mass	Relative isotopic abundance	Ionization potential (eV)
1	Hydrogen	H	1·00797	1	1·0078252	99·9850	13·595
				2	2·0141022	0·0149	
2	Helium	He	4·00260	3	3·0160297	0·0001	
				4	4·0026033	99·9999	24·6 ± 0·1
3	Lithium	Li	6·93900	6	6·0151234	7·4200	
				7	7·0160048	92·5800	5·390
4	Beryllium	Be	9·01220	9	9·0121828	100·0000	9·320
5	Boron	B	10·81100	10	10·0129385	19·7800	
				11	11·0093053	80·2200	8·296
6	Carbon	C	12·01115	12	12·0000000	98·8930	11·3
				13	13·0033551	1·1070	
7	Nitrogen	N	14·00670	14	14·0030744	99·6337	14·53
				15	15·0001093	0·3663	
8	Oxygen	O	15·99940	16	15·9949150	99·7590	13·64
				17	16·9991333	0·0374	
					17·9991600	0·2039	
9	Fluorine	F	18·99840	19	18·9984046	100·0000	17·418
10	Neon	Ne	20·18300	20	19·9924405	90·9200	21·8 ± 0·5
				21	20·9938474	0·2570	
				22	21·9913848	8·8200	
11	Sodium	Na	22·98980	23	22·9897703	100·0000	5·14
12	Magnesium	Mg	24·31300	24	23·9850443	78·7000	7·644
				25	24·9858385	10·1300	
				26	25·9825944	11·1700	
13	Aluminium	Al	26·98153	27	26·9815406	100·0000	5·984
14	Silicon	Si	28·08600	28	27·9769286	92·2100	8·149
				29	28·9764969	4·7000	
				30	29·9737722	3·0900	
15	Phosphorus	P	30·97380	31	30·9737633	100·0000	10·484
16	Sulphur	S	32·06400	32	31·9720728	95·0000	10·357
				33	32·9714591	0·7600	
				34	33·9678701	4·2200	
				36	35·9670791	0·0136	
17	Chlorine	Cl	35·45300	35	34·9688536	75·5290	13·01
				37	36·9659030	24·4710	
18	Argon	Ar	39·94800	36	35·9675465	0·3370	
				38	37·9627330	0·0630	
				40	39·9623842	99·6000	15·755
19	Potassium	K	39·10200	39	38·9637089	93·1000	4·339
				40	39·9640001	0·0118	
				41	40·9618270	6·8800	
20	Calcium	Ca	40·08000	40	39·9625921	96·9700	6·111
				42	41·9586281	0·6400	
				43	42·9587774	0·1460	
				44	43·9554875	2·0600	
				46	45·9536890	0·0035	
				48	47·9525260	0·1850	
21	Scandium	Sc	44·95592	45	44·9559174	100·0000	6·54

APPENDIX—*continued*

Atomic no.	Element	Symbol	Relative atomic mass	Mass no.	Precise mass	Relative isotopic abundance	Ionization potential (eV)
22	Titanium	Ti	47·90000	46	45·9526296	7·9300	
				47	46·9517670	7·2800	
				48	47·9479491	73·9400	6·82
				49	48·9478721	5·5100	
				50	49·9447843	5·3400	
23	Vanadium	V	50·94200	50	49·9471643	0·2400	
				51	50·9439644	99·7600	6·74
24	Chromium	Cr	51·99600	50	49·9460488	4·3100	
				52	51·9405102	83·7600	6·764
				53	52·9406510	9·5500	
				54	53·9388813	2·3800	
25	Manganese	Mn	54·93810	55	54·9380464	100·0000	7·432
26	Iron	Fe	55·84700	54	53·9396120	5·8200	
				56	55·9349339	91·6600	7·87
				57	56·9353907	2·1900	
				58	57·9332745	0·3300	
27	Cobalt	Co	58·93320	59	58·9331879	100·0000	7·86
28	Nickel	Ni	58·71000	58	57·9353358	67·8400	7·633
				60	59·9307795	26·2300	
				61	60·9310502	1·1900	
				62	61·9283396	3·6600	
				64	63·9279560	1·0800	
29	Copper	Cu	63·54000	63	62·9295898	69·0900	7·724
				65	64·9277890	30·9100	
30	Zinc	Zn	65·37000	64	63·9291400	48·8900	9·391
				66	65·9260395	27·8100	
				67	66·9271322	4·1100	
				68	67·9248481	18·5700	
				70	69·9253254	0·6200	
31	Gallium	Ga	69·72000	69	68·9255795	60·4000	6·00
				71	70·9247044	39·6000	
32	Germanium	Ge	72·59000	70	69·9242520	20·5200	
				72	71·9220823	27·4300	
				73	72·9234644	7·7600	
				74	73·9211786	36·5400	7·88
				76	75·9214042	7·7600	
33	Arsenic	As	74·92160	75	74·9216003	100·0000	9·81
34	Selenium	Se	78·96000	74	73·9224770	0·8700	
				76	75·9192117	9·0200	
				77	76·9199130	7·5800	
				78	77·9173093	23·5200	
				80	79·9165253	49·8200	9·75
				82	81·9167080	9·1900	
35	Bromine	Br	79·90900	79	79·9183320	50·5370	11·84
				81	80·9162920	49·4630	
36	Krypton	Kr	83·80000	78	77·9204010	0·3540	
				80	79·9163760	2·2700	
				82	81·9134820	11·5600	
				83	82·9141310	11·5500	
				84	83·9115053	56·9000	13·996
				86	85·9106160	17·3700	
37	Rubidium	Rb	85·47000	85	84·9117990	72·1500	4·126
				87	86·9091870	27·8500	
38	Strontium	Sr	87·62000	84	83·9134305	0·5600	
				86	85·9092763	9·8600	
				87	86·9088935	7·0200	
				88	87·9056283	82·5600	5·692

APPENDIX—continued

Atomic no.	Element	Symbol	Relative atomic mass	Mass no.	Precise mass	Relative isotopic abundance	Ionization potential (eV)
39	Yttrium	Y	88·90500	89	88·9058667	100·0000	6·38
40	Zirconium	Zr	91·22000	90	89·9047105	51·4600	6·84
				91	90·9056434	11·2300	
				92	91·9050386	17·1100	
				94	93·9063202	17·4000	
				96	95·9082920	2·8000	
41	Niobium	Nb	92·90600	93	92·9063803	100·0000	6·84
42	Molybdenum	Mo	95·94000	92	91·9068083	15·8400	
				94	93·9050891	9·0400	
				95	94·9058369	15·7200	
				96	95·9046748	16·5300	
				97	96·9060227	9·4600	
				98	97·9054100	23·7800	7·10
				100	99·9074775	9·6300	
43	Technetium	Tc					7·28
44	Ruthenium	Ru	101·07000	96	95·9075980	5·5100	
				98	97·9052890	1·8700	
				99	98·9059369	12·7200	
				100	99·9042173	12·6200	
				101	100·9055766	17·0700	
				102	101·9043481	31·6100	7·364
				104	103·9054280	18·5800	
45	Rhodium	Rh	102·90500	103	102·9055120	100·0000	7·46
46	Palladium	Pd	106·40000	102	101·9056070	0·9600	
				104	103·9040140	10·9700	
				105	104·9050860	22·2300	
				106	105·9034870	27·3300	8·33
				108	107·9038910	26·7100	
				110	109·9051640	11·8100	
47	Silver	Ag	107·87000	107	106·9050910	51·8200	7·574
				109	108·9047546	48·1800	
48	Cadmium	Cd	112·40000	106	105·9064630	1·2150	
				108	107·9041894	0·8750	
				110	109·9030101	12·3900	
				111	110·9041855	12·7500	
				112	111·9027628	24·0700	
				113	112·9044074	12·2600	
				114	113·9033668	28·8600	8·991
				116	115·9047615	7·5800	
49	Indium	In	114·82000	113	112·9040890	4·2800	
				115	114·9038750	95·7200	5·785
50	Tin	Sn	118·69000	112	111·9048340	0·9600	
				114	113·9027760	0·6600	
				115	114·9033530	0·3500	
				116	115·9017483	14·3000	
				117	116·9029606	7·6100	
				118	117·9016126	24·0300	
				119	118·9033159	8·5800	
				120	119·9022073	32·8500	7·342
				122	121·9034511	4·7200	
				124	123·9052830	5·9400	
51	Antimony	Sb	121·75000	121	120·9038223	57·2500	8·639
				123	122·9042203	42·7500	

APPENDIX—continued

Atomic no.	Element	Symbol	Relative atomic mass	Mass no.	Precise mass	Relative isotopic abundance	Ionization potential (eV)
52	Tellurium	Te	127·60000	120	119·9040240	0·0890	
				122	121·9030560	2·4600	
				123	122·9042817	0·8700	
				124	123·9028302	4·6100	
				125	124·9044263	6·9900	
				126	125·9033124	18·7100	
				128	127·9044675	31·7900	
				130	129·9062319	34·4800	9·01
53	Iodine	I	126·90440	127	126·9044755	100·0000	10·454
54	Xenon	Xe	131·30000	124	123·9061200	0·0960	
				126	125·9042790	0·0900	
				128	127·9035323	1·9190	
				129	128·9047840	26·4400	
				130	129·9035108	4·0800	
				131	130·9050840	21·1800	
				132	131·9041568	26·8900	12·127
				134	133·9053980	10·4400	
				136	135·9072220	8·8700	
55	Caesium	Cs	132·90500	133	132·9054360	100·0000	3·893
56	Barium	Ba	137·34000	130	129·9062840	0·1010	
				132	131·9050450	0·0970	
				134	133·9044930	2·4200	
				135	134·9056710	6·5900	
				136	135·9045590	7·8100	
				137	136·9058150	11·3200	
				138	137·9052350	71·6600	5·210
57	Lanthanum	La	138·91000	138	137·9071610	0·0890	
				139	138·9064030	99·9110	5·61
58	Cerium	Ce	140·12000	136	135·9071800	0·1930	
				138	137·9060270	0·2500	
				140	139·9054840	88·4800	6·5
				142	141·9093000	11·0700	
59	Praseodymium	Pr	140·90700	141	140·9076980	100·0000	5·7
60	Neodymium	Nd	144·24000	142	141·9077660	27·1100	5·7
				143	142·9098560	12·1700	
				144	143·9101290	23·8500	
				145	144·9126100	8·3000	
				146	145·9131530	17·2200	
				148	147·9169290	5·7300	
				150	149·9209210	5·6200	
61	Promethium	Pm		No stable isotopes			
62	Samarium	Sm	150·35000	144	143·9120740	3·0900	
				147	146·9149250	14·9700	
				148	147·9148510	11·2400	
				149	148·9172110	13·8300	
				150	149·9173030	7·44	
				152	151·9197550	26·7200	5·6
				154	153·9222220	22·7100	
63	Europium	Eu	151·96000	151	150·9198830	47·8200	
				153	152·9212600	52·1800	5·67
64	Gadolinium	Gd	157·25000	152	151·9198170	0·2000	
				154	153·9208910	2·1500	
				155	154·9226360	14·7300	
				156	155·9221430	20·4700	
				157	156·9239720	15·6800	
				158	157·9241230	24·8700	6·16
				160	159·9270710	21·9000	

APPENDIX—continued

Atomic no.	Element	Symbol	Relative atomic mass	Mass no.	Precise mass	Relative isotopic abundance	Ionization potential (eV)
65	Terbium	Tb	158·92400	159	158·9253860	100·0000	6·7
66	Dysprosium	Dy	162·50000	156	155·9243260	0·0524	
				158	157·9244400	0·0902	
				160	159·9252310	2·2940	
				161	160·9269700	18·8800	
				162	161·9268380	25·5300	
				163	162·9287700	24·9700	
				164	163·9292180	28·1800	6·8
67	Holmium	Ho	164·93000	165	164·9303570	100·0000	
68	Erbium	Er	167·26000	162	161·9288260	0·1360	
				164	163·9292350	1·5600	
				166	165·9303240	33·4100	
				167	166·9320790	22·9400	
				168	167·9324020	27·0700	
				170	169·9354910	14·8800	
69	Thulium	Tm	168·93400	169	168·9342450	100·0000	
70	Ytterbium	Yb	173·04000	168	167·9339250	0·1350	
				170	169·9347920	3·0300	
				171	170·9363540	14·3100	
				172	171·9364050	21·8200	
				173	172·9382340	16·1300	
				174	173·9388810	31·8400	6·2
				176	175·9425820	12·7300	
71	Lutetium	Lu	174·97000	175	174·9407960	97·4100	
				176	175·9427050	2·5900	
72	Hafnium	Hf	178·49000	174	173·9401400	0·1800	
				176	175·9414290	5·2000	
				177	176·9432450	18·5000	
				178	177·9437230	27·1400	
				179	178·9458400	13·7500	
				180	179·9465750	35·2400	7·0
73	Tantalum	Ta	180·94800	180	179·9475690	0·0123	
				181	180·9480280	99·9877	7·88
74	Tungsten	W	183·85000	180	179·9467000	0·1350	
				182	181·9482480	26·4100	
				183	182·9502660	14·4000	
				184	183·9509750	30·6400	7·98
				186	185·9544020	28·4100	
75	Rhenium	Re	186·20000	185	184·9530070	37·0700	
				187	186·9557910	62·9300	7·87
76	Osmium	Os	190·20000	184	183·9525950	0·0180	
				186	185·9538830	1·5900	
				187	186·9557880	1·6400	
				188	187·9558770	13·3000	
				189	188·9581830	16·1000	
				190	189·9584820	26·4000	
				192	191·9615140	41·0000	8·7
77	Iridium	Ir	192·20000	191	190·9606310	37·3000	
				193	192·9629640	62·7000	9·0
78	Platinum	Pt	195·09000	190	189·9599650	0·0127	
				192	191·9610780	0·7800	
				194	193·9627130	32·9000	
				195	194·9648040	33·8000	9·0
				196	195·9649650	25·3000	
				198	197·9678950	7·2100	
79	Gold	Au	196·96700	197	196·9665480	100·0000	9·22

APPENDIX—*continued*

Atomic no.	Element	Symbol	Relative atomic mass	Mass no.	Precise mass	Relative isotopic abundance	Ionization potential (eV)
80	Mercury	Hg	200·59000	196	196·9658220	0·1460	
				198	197·9667480	10·0200	
				199	198·9682750	16·8400	
				200	199·9683210	23·1300	
				201	200·9703040	13·2200	
				202	201·9706430	29·8000	10·43
				204	203·9734980	6·8500	
81	Thallium	Tl	204·3700	203	202·9723480	29·5000	
				205	204·9744380	70·5000	6·106
82	Lead	Pb	207·19000	204	203·9730490	1·4800	
				206	205·9744750	23·6000	
				207	206·9759030	22·6000	
				208	207·9766580	52·3000	7·415
83	Bismuth	Bi	208·98000	209	208·9804010	100·0000	7·287
84	Polonium	Po					8·43
85	Astatine	At					10·4
86	Radon	Rn					10·746
87	Francium	Fr					3·83
88	Radium	Ra					5·277
89	Actinium	Ac					6·9
90	Thorium	Th	232·03800	232	232·0380740	100·0000	
91	Protoactinium	Pa		No stable isotopes			
92	Uranium	U	238·04000	234	234·0409750	0·0056	
				235	235·0439440	0·7204	
				238	238·0508160	99·2739	6·08 ± 0·08

FURTHER READING

Reminiscence

Rutherford—Recollections of the Cambridge Days, M. Oliphant, Elsevier Publishing Co., Amsterdam, 1972.

Recollections and Reflections, J. J. Thomson, G. Bell and Sons, Ltd., London, 1936.

J. J. Thomson and the Cavendish Laboratory, G. P. Thomson, Thomas Nelson Ltd., London, 1964.

Early text books

The Discharge of Electricity Through Gases, J. J. Thomson, Westminster Archibald Constable and Co., U.S.A., 1898.

Conduction of Electricity Through Gases, J. J. Thomson, Cambridge University Press, 1903.

Rays of Positive Electricity, J. J. Thomson, Longmans, Green and Co., London, 1921.

Mass Spectra and Isotopes, F. W. Aston, Edward Arnold, London, 1933.

Mass Spectroscopy, H. E. Duckworth, Cambridge University Press, 1958.

Recent specialized texts and reviews

The Mass Spectra of Organic Molecules, J. H. Beynon, R. A. Saunders and A. E. Williams, Elsevier Publishing Co., Amsterdam, 1968.

Mass Spectrometry of Inorganic and Organometallic Compounds, M. R. Litzow and T. R. Spalding, Elsevier Publishing Co., Amsterdam, 1973.

Mass Spectrometry in Science and Technology, F. A. White, John Wiley and Sons, Inc., U.S.A., 1968.

Quadrupole Mass Spectrometry, P. H. Dawson, Elsevier Publishing Co., Amsterdam, 1976.

INDEX

Acetylacetone 71–73
Activators 126
Adsorption isotherm 58, 59
Age of the solar system 125
Alcoholates 40
Alkali metal halides 77
Amino acid sequence 68, 141
Anthracene-9,10-dione 54
Appearance potential 17, 45
Aromatic structure 49
Arrhenius equation 101
Arsine 83
Artificial intelligence 140
Aston, F. W. 6, 83, 145

Bainbridge, K. T. 13
Beer's Law 102
Belemnite 123
Benzo(e)pyrene 50
Biphenylene ion 54
Bleakney, W. 13
Blood gases 132
Body pool 135
Bond dissociation energy 44, 47
Boron hydrides 80
Burners 106
Butane-2,3-dione dioxime 75
Butanol 88

Caesium chloride 78
Calibration 115
Calutron 16
Carbon 79
Carrier gas 64
Cathode rays 1
Centroid 142
Charge exchange 103
Chelates 71, 73
Chemical ionization 38, 104
Chemi-ionization 116
Chemiluminescence 116
Cholesterol 20, 134
Chromatogram of halogenated alkanes 63
Chromatographic methods 58
Collision cross-sections 35, 101

Computer 137
Contact potential 46
Cracking pattern 57, 140
Craig, R. D. 19
Cross-section 35, 101
Cyclopentadienyl metal compounds 85
Cyclotron resonance mass spectrometers 26

Dating 125
Deconvolution 46
Deflection, in a magnetic field 2
  in an electric field 2
Dempster, A. J. 10, 91
Deuterium exchange 65
Diatomic molecule 42
Digitization 138
Diketone 73
Dimethylberyllium 78
Dinonyl phthalate 62
Direct insertion probe 20, 146
Direction focusing mass spectrometer 10
Dissociation 42
Dissociative charge transfer 103
Distribution coefficient 59
Double-beam mass spectrometer 2
Double collector 17
Double focusing ix, 17
Doubly charged ion 54
Drift tube 25
Drugs 133

Ebulliometry 50
Einstein, A. 145
Electron bombardment ion sour 16, 34
  impact method 47
  monochromator 46
  multiplier 14
Energetics of ionization 42
Enthalpy of formation 47
Ephedrine 38, 39
Errock, G. A. 19
Ethyllithium 78
Ethylpotassium 78

Far ultraviolet 36
Ferrocene 85
Field ionization x, 39
 source 50
Fingerprint 56
Fischer O. 33
Fite, W. L. 30
Flame ionization detector 62, 117
Flames 104
Fortuitine 68
Fractionation of isotopes 123
Fragment ion peaks 52
Franck–Condon principle 43, 48
Fringe field 31

Gas chromatography 35, 61
Geochronology 122
Geology 121
Glycerides 65
Glycerol 65
Glucose 40
Goldstein, E. 1

Half-life 128
Hall effect 118
Halogen compounds 89
Helium 123
 discharge 37
Hemispherical condenser 38
Herzog, R. F. 19, 91, 131
Hexafluorobenzene 50
Hipple, J. A. 16
Homologous series 49
Hydrates 40
Hydrocarbons, polycyclic 50, 60
8-Hydroxyquinoline 74

Integrated Ion Current method 143
Interatomic distance 42
Ion atmospheres 131
 kinetic energy 23
 –molecule reactions 96
 storage 34
 trajectories 28
Ions in flames 116
Ionization efficiency 142
 curve 46
 energy 44
 potential 17, 47
Isomers of metal chelates 73
Isoprene 67
Isotope dilution method 135
 effect 124
 fractionation 123
 ratios 51

Isotopes, discovery of 6
 electromagnetic separation of 16
Isotopic abundance 13
 determination of by double collector 17
 peak 50

Jet separators 64
Johnson, W. H. 20

Kinetic energy 42
Knudsen cell 70, 71

Langevin, P. 102
Lithium fluoride 36, 78
 hydroxide 78

Macrocyclic pigments 69
Magnetic sector 14
Martin, A. J. P. 61, 62
Mass filter 27, 28
 fragmentography 143
 marking 21, 139
 range 145
 spectrograph 4
  of Aston, F. W. 7
 spectrometer, in geology 13
 spectrum, first 4, 5
  of residual gases 5
 synchrometer 26, 144
Matsuda, H. 144
Mattauch, J. 19, 91, 131
Maxwell–Boltzmann distribution 145
Membrane separator 64
Mercury (II) chloride 52
Metal carbonyls 83
 chelates 71
Metastable peak xi, 55
Meteorites 128
Methyl trideuteromethyl sulphide 54
Microdensitometer 93
Mobility, of ions 118
Molecular beam 100, 111
Monopole 31
 mass filter 33
Morgan, G. 71

Negative ions 86
Neon, mass spectrum of 6
Neutron pair separation energy 145
Nickel chelates 75
Nier, A. O. C. 14, 20, 121, 123, 127
Noble gases 89, 98

Ohm's Law 118
Organometallic compounds 83

158

Packing fraction 9
  curve 9, 145
Palladium 75
Paper chromatography 59
Parabola 2, 3
  of residual gas ions 3
Parent ion peak 49
Parkinson's disease 136
Partial pressure analyser 32
Paul, W. 27, 33
Peak matching 21
Peptide 67
Peptidolipid 68
Perchloryl fluoride 86
Perfluoroacylation of amines 41, 42
Perfluorocyclohexane 53
Phosphine 83
Photoelectron spectrometry 37
  spectrum 38
Photoionization 36
Plastoquinones 66
Plumbane 82
Pneumatic probe 119
Polycyclic hydrocarbons 50, 60
Polyisotopic nature of carbon 50, 51
  halogens 51
  sulphur 51
Positive ray 1
  analyser 2
Potential energy 42, 87
Precise mass 22
Primeval isotope ratio 127
Propane-1,2,3-triol (glycerol) 65
Propanone 95
Proton affinity 39
  transfer 103, 104

Quadrupole field 27
  mass filter 30
Quasi-equilibrium theory 48
Quistor 33

Radio-frequency field 27
Radiolysis 97
Raoult's Law 61
Rearrangement ion 54
Residual gas analysers 30
Resolution 6, 144
  definition of 12
  of Aston's mass spectrograph 8
  variation with slit width 12
Respiration 132
Reynold's number 119
Ruthenium 71

Salicylaldoxime 75
Sample introduction 17
Scavenger 108
  probe 108
Schizophrenia 135
Selenium 79
Sensitivity 142
Silane 82
Silicon carbide 79
Smith, L. G. 27
Soddy, F. 6
Solar wind 130
Sonic flow 119
Spark source mass spectrograph 70, 91, 132
Spectrum matching 56, 57
Spodumene 123
Sputtering 129
Steinwedel, H. 33
Steroids 133
Sulphur 79

Tetraethyllead 82
Tetrafluoromethane 49
Thermal ionization 11, 40, 126
Thermionic emission 34
Thin-layer chromatography 58–60
Thomson, J. J. 1, 83, 96
Time-of-flight mass spectrometer 25
Trap electrode 34, 35
Trimethylsilyl ethers 42, 146
Tungsten filament 34
p-Tyramine 136

Ultraviolet, far 36
Ultraviolet galvanometer recorder 21
Unimolecular decomposition 48
Uranium–lead system 127
Urea 135

Velocity, of ions 6
Vertical ionization potential 44
Volatility 41
Von Zahn, U. 27

Water, temperature of 123

Xenon fluorides 90
  oxyfluorides 90

THE WYKEHAM SCIENCE SERIES

1	*Elementary Science of Metals*	J. W. Martin and R. A. Hull
2	†*Neutron Physics*	G. E. Bacon and G. R. Noakes
3	†*Essentials of Meteorology*	D. H. McIntosh, A. S. Thom and V. T. Saunders
4	*Nuclear Fusion*	H. R. Hulme and A. McB. Collieu
5	*Water Waves*	N. F. Barber and G. Ghey
6	*Gravity and the Earth*	A. H. Cook and V. T. Saunders
7	*Relativity and High Energy Physics*	W. G. V. Rosser and R. K. McCulloch
8	*The Method of Science*	R. Harré and D. G. F. Eastwood
9	†*Introduction to Polymer Science*	L. R. G. Treloar and W. F. Archenhold
10	†*The Stars; their structure and evolution*	R. J. Tayler and A. S. Everest
11	*Superconductivity*	A. W. B. Taylor and G. R. Noakes
12	*Neutrinos*	G. M. Lewis and G. A. Wheatley
13	*Crystals and X-rays*	H. S. Lipson and R. M. Lee
14	†*Biological Effects of Radiation*	J. E. Coggle and G. R. Noakes
15	*Units and Standards for Electromagnetism*	P. Vigoureux and R. A. R. Tricker
16	*The Inert Gases: Model Systems for Science*	B. L. Smith and J. P. Webb
17	*Thin Films*	K. D. Leaver, B. N. Chapman and H. T. Richards
18	*Elementary Experiments with Lasers*	G. Wright and G. Foxcroft
19	†*Production, Pollution, Protection*	W. B. Yapp and M. I. Smith
20	*Solid State Electronic Devices*	D. V. Morgan, M. J. Howes and J. Sutcliffe
21	*Strong Materials*	J. W. Martin and R. A. Hull
22	†*Elementary Quantum Mechanics*	Sir Nevill Mott and M. Berry
23	*The Origin of the Chemical Elements*	R. J. Tayler and A. S. Everest
24	*The Physical Properties of Glass*	D. G. Holloway and D. A. Tawney
25	*Amphibians*	J. F. D. Frazer and O. H. Frazer
26	*The Senses of Animals*	E. T. Burtt and A. Pringle
27	†*Temperature Regulation*	S. A. Richards and P. S. Fielden
28	†*Chemical Engineering in Practice*	G. Nonhebel and M. Berry
29	†*An Introduction to Electrochemical Science*	J. O'M. Bockris, N. Bonciocat, F. Gutmann and M. Berry
30	*Vertebrate Hard Tissues*	L. B. Halstead and R. Hill
31	†*The Astronomical Telescope*	B. V. Barlow and A. S. Everest
32	*Computers in Biology*	J. A. Nelder and R. D. Kime
33	*Electron Microscopy and Analysis*	P. J. Goodhew and L. E. Cartwright
34	*Introduction to Modern Microscopy*	H. N. Southworth and R. A. Hull
35	*Real Solids and Radiation*	A. E. Hughes, D. Pooley and B. Woolnough
36	*The Aerospace Environment*	T. Beer and M. D. Kucherawy
37	*The Liquid Phase*	D. H. Trevena and R. J. Cooke
38	†*From Single Cells to Plants*	E. Thomas, M. R. Davey and J. I. Williams
39	*The Control of Technology*	D. Elliott and R. Elliott
40	*Cosmic Rays*	J. G. Wilson and G. E. Perry
41	*Global Geology*	M. A. Khan and B. Matthews
42	†*Running, Walking and Jumping: The science of locomotion*	A. I. Dagg and A. James
43	†*Geology of the Moon*	J. E. Guest, R. Greeley and E. Hay
44	†*The Mass Spectrometer*	J. R. Majer and M. P. Berry
45	†*The Structure of Planets*	G. H. A. Cole and W. G. Watton
46	†*Images*	C. A. Taylor and G. E. Foxcroft
47	†*The Covalent Bond*	H. S. Pickering
48	†*Science with Pocket Calculators*	D. Green and J. Lewis

THE WYKEHAM ENGINEERING AND TECHNOLOGY SERIES

1	*Frequency Conversion*	J. Thomson, W. E. Turk and M. J. Beesley
2	*Electrical Measuring Instruments*	E. Handscombe
3	*Industrial Radiology Techniques*	R. Halmshaw
4	*Understanding and Measuring Vibrations*	R. H. Wallace
5	*Introduction to Tribology*	J. Halling and W. E. W. Smith

All orders and requests for inspection copies should be sent to the appropriate agents. A list of agents and their territories is given on the verso of the title page of this book.

†(*Paper and Cloth Editions available.*)